Whatever happened to EDEN?

Earth's Energy-Environment Crisis Opens Doors to New Prosperity

John R. Sheaffer
and
Raymond H. Brand

Tyndale House
Publishers, Inc.
Wheaton, Illinois

Library of Congress
Catalog Card Number 80-50665
ISBN 0-8423-7871-5, paper
Copyright © 1980
by John R. Sheaffer and
Raymond H. Brand. All rights reserved.
First printing, October 1980.
Printed in the
United States of America.

CONTENTS

PREFACE

Earth is running out of energy, vital raw materials, and living space for people. More important, we are running out of time to correct these serious deficiencies. Already, environmental scientists have detected in our earth ominous signs, symptoms akin to human problems: hardening of the arteries, advanced emphysema in the lungs, and the senile pallor of internal decay. Air, water, and the soil of planet Earth are being stripped of their vitality by overburdening use and under-nourishing replenishment. Presently supporting more than four billion inhabitants, Earth's life-systems require suffi-cient oxygen, water, and circulating nutrients to remain pro-ductively alive. Now, that life is being choked by humanity's contemptuous, ignorant disregard of Earth's care and feeding.

Some environmentalists have pointed a finger at technol-ogy and have mounted a crusade against it to save the human race from extinction. As a scientist-business executive and a college ecology professor, the authors agree that human survi-val is eminently desirable. But we call for a whole new quality of life that is based on managing technology and using it to replenish the environment. Humanity possesses productive ingenuity and Earth contains the rich resources. We urge a fruitful partnership of the two, characterized by mutual service.

Responsible management of Earth's resources is a moral

and spiritual issue. We believe every person is a unique creation of God and is responsible for the welfare of his neighbor,
as well as for the environmental legacy that is handed down to
the next generation. We do not have the right to waste the
world's resources while less-advantaged peoples struggle to
survive on the leftovers. Although the adoption of simpler
life-styles in highly industrialized nations would help somewhat, much more would be conserved and accomplished in
our society if we practiced efficiency in our personal and
public activities. Prodigal waste and callous exploitation
have rushed us to the brink of economic, energy, and environmental crisis. But history, experience, and our faith in a re-
creating, redeeming God convince us that highly motivated
men and women can change these imminent dangers into
grand opportunities.

For several years the authors of this book have joined in
related pursuits. We live in the same town, serve in the same
church, have campaigned together in politics, shared victories
and defeats in the professional world, and attended the wed-
dings of each other's children. Foremost in this present endeav-
or is our desire to inform conscientious people about the
causes and depth of the energy-environment peril and to de-
scribe personal actions and public policies that can revive
Earth and its tenants. This effort is part of our personal
stewardship of resources, a stewardship that is one of our
responsibilities to God the Creator.

We express our deep gratitude for the editorial collaboration of Ted Miller on this project.

Jack Sheaffer
Ray Brand

ONE
WHAT HAPPENED
TO UTOPIA?

Strains of Bach's fugue on a Moog synthesizer danced from the clock radio into the bedroom, stirring one of the slumbering figures in the four-poster. As if by habit, a well-manicured hand appeared from beneath the electric blanket and grasped a chenille robe from the bedside chair. Without looking, Celeste Grey knew the digital clock read 6:30 and it was time to launch her family into another day in Sacramento, California.

First on the schedule was plugging in the coffee maker. Peter could scarcely see without his morning coffee, and it had to be ready at 6:45.

Celeste underlined last night's mental note to replenish her coffee supply—three dollars, four dollars a pound, the price was irrelevant though certainly not immaterial. On the way to the bathroom she flipped the furnace thermostat to remove the morning chill. Northern California needed both winter heating and summer cooling to keep its residents comfortable.

The telephone caught Celeste in midstep. Who could be . . . ? "Oh, Henry, no! That means . . . yes, right away. Good-bye." Mumbling to herself, Celeste hurried to the bedroom and shook Peter. "Honey, get up! Henry's car won't start and he says you have to drive to Frisco today. You've got twenty-three minutes to get ready."

Peter's groan trailed after Celeste as she darted down the hall. The next hour became a blur of toothbrushes, blender whirr, clothes sorting, cold toast, TV blare, steamy bathrooms, squealing tires, spilled cereal, squabbling children, a blown electrical fuse, hasty hugs, and unrecognizable noises and words before Peter was dispatched and Kevin, Robin, and Todd were packed in the family station wagon for their ride to junior high, elementary, and nursery schools. Just in time Celeste remembered it was trash day, and the four of them managed to wrestle three heavy garbage containers to the curb and still arrive at school on time.

A fairly typical morning for the mythical Grey family and also for real-life, middle-class Americans in the eighties. It features full schedules, overloaded energy circuits, ready mobility, and throwaway products. Celeste and her flesh-and-blood counterparts would hardly count themselves citizens of Utopia, but they are satisfied enough with their status to struggle for more of the same, and they fully expect today's situation to improve. Isn't that what stock options, home-owners associations, women's groups, labor unions, and political parties are all about? And modern prophets of civilization have encouraged their bright expectations.

THE MAGICAL A.D. 2000

Science and technology confidently pointed Americans toward more and better of everything after World War II. Scanning the future horizons of electronics in 1955, the chairman of the Radio Corporation of America, David Sarnoff, asserted: "There is no element of material progress we know today that will not seem, from the vantage point of 1980, a fumbling prelude. There is no longer margin for doubt that whatever the mind of man visualizes, the genius of modern science can turn into fact."[1]

Analyzing, restructuring, and programming, science-based technology planned a new world for mankind. Housewives got a glimpse of a penthouse paradise from a British engineering professor in 1964. Meredith Thring forecast: "Within ten to twenty years' time we could have a robot that will completely eliminate all routine operations around the house and remove the drudgery from human life. It will be a machine having no more emotions than a car, but having a memory for instructions and a limited degree of instructed or built-in adaptability according to the positions in which it finds various types of objects. The mechanical helper will operate other more specialized machines, for example, the vacuum cleaner or clothes washing machine."[2]

New airborne vehicles, major reductions in hereditary defects, effective appetite and weight control, programmed relaxation and sleep, automated stores, low-cost homes, inexpensive worldwide transportation, and artificial moons for lighting large areas at night were predicted as very likely by the end of the twenty-first century, according to Herman Kahn and Anthony Wiener. Worldwide, these seers forecast intensive/extensive expansion of tropical agriculture and forestry, improved uses of the oceans for mining and "farming," and global use of nuclear reactors for power.[3]

Looking ahead half a century, leaders in transportation, communications, education, and medicine anticipated a world of speed, comfort, knowledge, and health. They still visualized gulfs between the rich and the poor of earth, but they promised technological marvels and unimagined abundance.

Najeeb E. Halaby, former administrator of the Federal Aviation Agency, set the stratosphere as the limit for future air transportation. Passenger airliners can be built as large as is economically feasible and as fast as is socially desirable, he declared. J. R. Pierce, electronic researcher, foresaw quick, cheap, long-distance communication that is "more like meeting people face to face" than mere vocal telephone contacts.

Homes around the world will be educational centers when technology develops low-cost equipment for sending and receiving information. Mathematician and computer specialist Anthony Dettinger declares that "the instant university is possible," with printed and filmed information and events accessible by pushing the right buttons.[4]

New people as well as new things are in the offing if the sensational scenarios of science writer Gordon Rattray Taylor come true. In *The Biological Time Bomb*, Taylor forecast such wonders as routine transplants of human organs, drug treatment that expands awareness or lowers anti-social tendencies, parental choices of children's gender, and eventual "carbon copying" of model people through cloning.

The dazzling scope of science's revolution was projected by Taylor: "We can now create, on a commercial scale by chemical processes, substances which previously we had to look for in nature, and even substances which never previously existed. Whereas before we had to make do with what nature provided, now we can decide what we want; this may be called chemical control. Similarly, in the coming century, we shall achieve biological control: the power to say how much life, or what sort, shall exist where. We shall even be able to create forms of life which never existed before."[5]

Experimenting with anti-aging cells in 1976, the Geriatric Research Center of the Veterans Administration Hospital in Los Angeles offered hope that young "t-" blood cells can be given to old people to help them resist such old-age diseases as cancer, arthritis, diabetes, and kidney infection. The vitality of these white blood cells in the body declines in advanced age, lowering one's natural immunity to invading viruses. Laboratory work by Dr. Takashe Makinodan has shown that the potent t- cells of young mice can restore immunity in older mice. Clinical tests on humans will determine how effective this new disease-fighting method may be.[6]

But trailing after these tantalizing previews of a technological Utopia came reverberations of danger. A new breed of scientist detected a death-shadow lurking over civilization and sounded the alarm. Though controversy swirled over the accumulating evidence, many scientists began to warn humanity that racial survival, not Utopian arrival, is the paramount issue of the future.

THE SCIENCE SLEUTHS

Marine biologist Rachel Carson was one of the first to report the footprints of synthetic death invading earth's life-systems. In 1962 her book *Silent Spring* warned that chemical pesticides were killings birds and fish as well as their intended targets: crop-devouring insects. Worse, some of the pesticides, such as DDT, were contaminating food for humans and accumulating in human milk with potentially dangerous effects.

These threatening findings were widely ignored. Pesticides had dramatically escalated food production—at least temporarily—to help meet the needs of an exploding world population. Soon afterward, other biologists joined the hunt, and early in the seventies their expose of extensive pollution of air, water, and soil made global headlines.

Barry Commoner, a Washington University biologist, alerted others to the significance of air pollution in 1953 when test explosions of nuclear bombs sprayed widely separated parts of the earth with radioactive fallout. Though the amounts were slight, an ominous result was transmission of radioactive strontium 90 through plants and milk deep into the bodies of humans. It lodged next to the bone marrow that produces blood cells, emitting beta rays which in large doses have caused cancer in test animals.

Anxious scientists and citizen groups launched an informational campaign that helped consummate a 1963 international treaty against nuclear testing in the atmosphere. Some nuclear powers refused to sign the treaty and continued to test bombs, but environmentalists applauded the step in the right direction.

The identity of the environmental villain became clearer when the Los Angeles smog mystery was solved. Industrial smokestacks and trash incinerators had belched sulfur dioxide into the atmosphere during the war production years of the early forties. Capping the stacks with dust-precipitators had thinned the grimy dustfall over the city, but Angelinos still squinted and coughed in a yellow-brown haze during the early fifties.

Health officials knew that a choking combination of smoke and fog in London, England, had killed 4,000 residents in a five-day siege, but L.A.'s plague was descending on sunny days. Seeking the cause, Dr. Haagen-Smit of the California Institute of Technology eventually determined that hydrocarbons and nitrogen oxides pouring from vehicle exhausts and petroleum refineries were interacting in the sunlight to produce the noxious haze which irritated eyes and lungs and sometimes disintegrated fabrics. This was "photochemical smog," and Los Angeles officials began a long struggle against its main producers, the auto industry.

Under government pressure based on the Clean Air Act of 1970, automakers gradually modified their four-wheeled pollution machines. Finally, twenty-five years after the problem was first recognized, car manufacturers produced vehicles that burn lead-free gasoline and carry exhaust devices that reduce the hydrocarbons, carbon monoxide, and nitrogen oxides spewed into the air. Without these government regulations, industry would have continued to poison the atmosphere and the humans who breathe it. Though many of the toxic effects are gradual, confirmed cases of illnesses and death from highly concentrated pollutants have proved the deadliness of the new chemical-technological civilization.

Steadily rising pollution of air, water, and soil pressed a new reality upon ecologists. Unless the chemical and thermal wastes of modern technology are checked and recycled into the natural systems which support plant, animal, and human life, the ingenious designer of today's civilization—mankind—will smother under his own technological creations. The signs are crystal clear to ecologists.

Writing in *The Closing Circle* in 1971, Barry Commoner said: "My own judgment, based on the evidence now at hand, is that the present course of environmental degradation, at least in industrialized countries, represents a challenge to essential ecological systems that is so serious that, if continued, it will destroy the capability of the environment to support a reasonably civilized human society. Some number of human beings might well survive such a catastrophe, for the collapse of civilization would reduce the pace of environmental degradation. What would then remain would be a kind of neobarbarism with a highly uncertain future."[7]

While admitting that no one can chart a timetable on the rate of deterioration or the date of ecological collapse, Commoner declares firmly: "We are in an environmental crisis because the means by which we use the ecosphere to produce wealth are destructive of the ecosphere itself. The present system of production is self-destructive; the present course of human civilization is suicidal."[8]

Nobel prize-winning microbiologist Rene Dubos and renowned economist Barbara Ward seconded that appraisal in their 1972 book, *Only One Earth*. "This is the hinge of history at which we stand, the door of the future opening on to a crisis more sudden, more global, more inescapable and more bewil-

14

dering than any ever encountered by the human species and one which will take decisive shape within the life span of children who are already born."[9]

The statistical report of the Club of Rome, an international group of businessmen and scientists who sponsored a computer evaluation of mankind's environmental and energy prospects, was more analytical. A research team headed by Massachusetts Institute of Technology professor Dennis Meadows fed computers with data on the trends of pollution, population growth, food production, industrialization, and consumption of resources. The result, published in *The Limits To Growth* in 1972, forecast global collapse unless living standards are radically lowered and world population is sharply curbed.[10]

"The crux of the matter," wrote the Club of Rome Executive Committee, "is not only whether the human species will survive, but even more whether it can survive without falling into a state of worthless existence." A later report sponsored by the same group concluded that moderate technological growth in developing areas along with coordinated planning could mitigate some of the predicted perils.[11]

Paul Ehrlich, the biologist who warned against catastrophe from unrestrained population growth in *The Population Bomb* back in 1968, wrote somewhat more hopefully five years later. In *Human Ecology*, he said: "If man changes his ways in time, turning away from folly and toward survival, civilization may endure the critical decades ahead . . . . Supporting a constant world population smaller than today's at a material standard of living lower than that now enjoyed in the United States would still require constant vigilance . . . . Human ecology—the study of human relationships with the environment—will persist as a discipline for a long time, if civilization is to persist for a long time."[12]

Instead of fashioning a technological paradise, some ecologists claim that the wizards of modern industry are crafting a global gas chamber. Unchecked pollution could destroy life when Earth's natural life systems are choked beyond recovery with technological and chemical wastes.

Until recently, manufacturers and consumers were unwitting accomplices in the throttling of plant Earth. But evidence amassed by a number of scientists now confirms that wastes discharged unimpeded into the air and water must be acknowledged as dire offenses against nature and humanity. The only uncertain factor is how quickly and widely nature will retaliate against its violators.

The genius of science and technology has boomeranged. Life-nourishing chemicals transplanted to incompatible environments have become death-dealing toxins. Atmospheric discharges from power plants, factories, and vehicles poison the air we breathe and the water we drink. Home sewage and industrial waste pumped into waterways nourish excessive plant life and terminate animal life. Though government regulations seek to restore clean air and water to the nation, powerful social and economic forces continue to resist environmental constraints.

An example of the technology-environment clash is the auto industry's approach to air emission control standards. Rather than redesign engines to make them more efficient and thereby reduce emissions, the industry first added exhaust emission equipment to reduce the amount of noxious hydrocarbons and carbon monoxide entering the atmosphere. The controls reduced emissions but also lowered fuel economy, forcing motorists to use more gas and canceling the advantage of the new equipment.

Aiming for more efficiency and less pollution, auto designers next increased the engine airtake to permit better combustion of gases. The disconcerting result was a greater expulsion of nitric oxide into the air from hotter explosions. Attacking this new threat with catalytic converters, the auto industry hailed its success until research showed that high-octane lead additives in gasoline were breaking down the catalysts after brief use.

Such piecemeal measures perpetuate inefficiency and blight while exacting high financial penalties. Everyone loses in the long run.

The same tug of war rages over clean water supplies and cheap consumer products. The steel, paper, plastics, rubber, petroleum, chemical, and food industries pumped huge quantities of waste into rivers and lakes until restricted by government decrees. The wastes choked aquatic life and increasingly affected all living things. But because installation of pollution controls boosted manufacturing costs, both producers and consumers resisted the health measures. In 1973 product prices and human well-being met head-on when fuel costs suddenly skyrocketed.

THE ENERGY CRISIS

With pollution and population growth shrinking both resources and living space, a gigantic crisis loomed over the world in 1973 as fuel for heating homes, powering factories, and propelling vehicles suddenly dwindled. Hints of future shortage had appeared in technical publications for twenty years. These forecasts became real in the winter of '72 as U.S. schools and factories closed temporarily for lack of heating fuel. The following summer, brownouts of electricity skipped across the nation as power demands exceeded supply. Although huge reserves of oil and coal still awaited development, rising costs of extraction and transport discouraged rapid expansion. Also, the pollution caused by use of high-sulfur coal put a damper on a significant portion of the reserves. Environmental engineer and government official S. David Freeman focused the dilemma: "Environmental goals and energy demands are on a collision course."[13]

The global jolt hit in late 1973. Oil-producing Arab countries stopped shipments of petroleum to nations that traded with Israel after the Arab alliance lost the Yom Kippur War. Nations quickly checked their oil supplies first and their ethical stance second and decided whether fuel at home or friendship with Israel was more important. No sooner were international loyalties realigned than economic shock waves hit the world money market: Arab oil producers joined other members of the Organization of Petroleum Exporting Countries (OPEC) in quadrupling the price of oil. With one massive stroke, the oil cartel turned the rich nations into debtors and the poor nations into beggars. As Paul Ehrlich wrote, "The OPEC revolution has transformed the power structure of the world dramatically and perhaps permanently."[14]

The impact of ransom-size oil prices on world economics was summarized by Alvin Toffler, author of *Future Shock*, in his later book *The Eco-spasm Report.* "Never before in the history of the global money system has one group of nations so swiftly, and with such surgical precision, captured so rich a monetary booty, and so destabilized the whole system. The problem, says Gary L. Seevers, a member of the Council of Economic Advisors in Washington, 'is something economists would not have dreamed of five or ten years ago.'"[15]

Fiscal nightmares haunted both rich nations and poor. Heavily industrialized nations slowed their economies, but the consequent unemployment and inflation induced many of them to stimulate economic activity at the higher prices. It has been estimated that the oil cartel will receive $200 billion for their product in 1979 compared to $18 billion in 1972.[16]

Meanwhile developing countries were forced to cut back on petroleum-based products, notably fertilizer for essential food production. Nevertheless, their expenditures for imported fuel jumped $15 billion in one year, and payments for fertilizer and food went up more than $5 billion.[17]

Inflated oil prices undermined the so-called Green Revolution in developing countries. A long, heartening struggle to breed higher-yielding strains of rice and wheat, the Green Revolution overcame governmental inertia and farmers' inexperience after improved hybrids were developed for particular areas of the earth. However, when fertilizer prices soared, many peasant farmers simply gave up the attempt to increase their food supply.

In the Philippine Islands, the first Asian nation to experiment with the Green Revolution, large fertilizer requirements and floods conspired to curtail the project. Nicholas Wade summarized the overall picture: "Enthusiasm about the Green Revolution has slowly subsided, dipping occasionally into positive vilification. That the Green Revolution 'was oversold,' 'is a myth,' 'will turn into a red revolution,' are some of the verdicts heard today."[18]

Striving to raise its food production through increased irrigation and fertilization, India's underfed millions staggered from the financial assault. Whereas India's oil bill was $140 million in 1970, the 1974 cost exceeded $1 *billion*—equaling the entire fiscal assistance pledged to it by other countries.

The cost of energy claimed 80 percent of India's total export receipts for the year.[19]

Energy prices are a critical factor in the ability of food-rich nations to meet the needs of the world's expanding population. David Pimentel, agricultural scientist at Cornell University, estimates that America's 240 percent increase in corn production between 1946 and 1970 required a 310 percent boost in energy use on the farms. Thus, despite increased production, the ratio of energy output (as measured in corn) to energy input (in petroleum and fertilizer) decreased substantially (see Figure 1).[20]

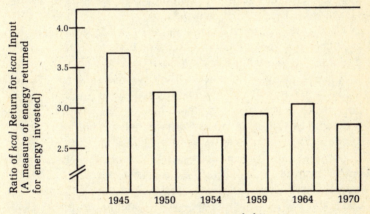

Declining ratio of energy yield per energy expended in corn production for selected years from 1945 to 1970.
(Data from D. Pimentel in *Science.*)

High fuel costs set off repercussions in every area of society. The prestigious University of Chicago initiated a $35 million bond issue for renovating its heating system and improving insulation of campus buildings after its energy bill tripled in 1976 over the 1972 cost. Little Geneva College in Beaver Falls, Pennsylvania, along with some other small northern schools, sent students home in midwinter of 1977 rather than pay a surcharge of thousands of dollars for gas heat when fuel became short. Hospitals and other public institutions added huge fuel increments to already burdensome costs in the hard winters of 1977 and 1978.

One of the most disquieting signs in the energy crisis was the American public's blase response. Though fuel shortages and mass layoffs spread gloom in 1974 and 1975, massive government expenditures spurred the economy and restored the traditional buy-now-pay-later buoyancy. Auto sales approached record totals in 1977 and high-powered marketing helped boost private debt to the $200 billion mark. The nation was flying high again—on IOUs.

In 1977 Americans bought (or charged) $26 billion more from other nations than they sold abroad. A large portion of that trade debt stemmed from oil purchases. America could not afford all the energy it consumed, yet it steadily imported more for its insatiable machines.

Who will pay the national fuel bill? International agencies are pondering that question, including European leaders who urge the U.S. to responsibly handle its energy debt and stabilize the shrinking dollar on world money markets. In 1978 the trade deficit increased to more than $28 billion, yet OPEC boosted oil prices again in 1979.

Former Secretary of State Henry Kissinger pointed out the dangers of unrestrained energy consumption in a June 11, 1977 news conference in Chicago. "We saw in 1973 what a small cut in imported oil can do," he told newsmen. "It caused the worst recession and highest unemployment and inflation in our history. The U.S. must reduce its demand for imported oil to maintain economic stability and an independent foreign policy."

Dean Rusk, professor of international law at the University of Georgia and former Secretary of State, tied energy needs to war outbreaks in an interview with U.S. News editors in Washington, D.C. "There is a new cause of war coming down the road. James Schlesinger said recently that oil and gas are going to run out in thirty to forty years. Long before that happens, nations of the world are going to be at each other's throats for energy supplies."

Like pollution threats, the energy shortage appears to many Americans as only a remote danger—a problem to be deferred until we actually are gagging on dirty air or bicycling to the unemployment office to pick up government handouts. By then it will be too late to turn things around.

Great Britain may herald its economic salvation in a fabulous North Sea oil strike, and water-short Saudi Arabia may offer $90 million from its bursting oil treasury for shippers to tow an iceberg 5,000 miles to its parched kingdom. But such limited measures will not resolve the worldwide resource and pollution crisis. Profligate use of energy by the industrial nations jeopardizes the economic well-being of developing nations that are linked with them in trade, and technological wastes imperil the health of people around the globe who subsist on the same polluted water and air.

THE WORLDWIDE EMERGENCY

Inch by inch, the technological era is shrinking our globe into an international community. National boundaries, different languages, and ideological conflicts still inhibit understanding and unity, but increasing travel, trade, and transmission of information are linking nations into closer partnerships. No modern nation is self-sufficient, and no longer can world leaders impose their wills arbitrarily on the small nations.

The magnetic pull of international associations is felt on many fronts. Nearly eight million Americans traveled abroad in 1976, and over four million visitors entered the United States. Trade among free-world nations rocketed from $56 billion a year to $875 billion in the same period. Investment in foreign businesses also escalated dramatically. U.S. investments shot up over 1,000 percent in twenty-five years to $133 billion, while foreign investors put $26 billion in U.S. operations.[21]

With multiplying links, the nations of the world increasingly depend on one another for economic well-being. But the supply of food and shelter and conveniences depend on nature's resources and the life-sustaining environment. Pollution in one area of the world affects health in other areas. Even where industrial wastes have not yet contaminated the air and water, the excess of population over food supply in certain locations causes calamities that humane neighbors cannot ignore.

"The whole world is going through a crisis today," declares journalist Andre Fontaine. "No state can claim to have been left wholly untouched by it. Monetary instability, inflation, and erratic commodity prices are hitting the poor nations just as badly as the industrial countries. Not even the Communist countries, forced as much out of political as of economic necessity to increase their trade with the West, have been spared the repercussions."[22]

Industrial pollution in Russia has reached a serious stage. "At the center of many of Russia's largest cities, the content of sulfur dioxide is two to two and one-half times greater than the maximum permissible concentration," according to John Kramer. Water pollution has also hit Russian waterways such as the Volga River. It has been estimated by the USSR Academy of Sciences that as much as 100 million cubic meters (264.2 billion gallons) of raw sewage enters Russian waterways each day.[23]

In Port-Au-Prince, capital of Haiti, residents scramble to escape floodwaters in heavy rainy seasons and search for drinking water during dry periods. The seasonal disasters flow from two hundred years of environmental depredations by the island's inhabitants, who now have the highest density population and lowest per capita income in the Western Hemisphere.

Although the island receives substantial rainfall, Haiti suffers from electricity deficiency, crop failures, diminishing drinking water, and severe malnutrition. Much of the blight is traceable to the island's wholesale deforestation by islanders in need of fuel and saleable products. As the trees vanished, topsoil eroded and rainfall accelerated its rush to the sea, decreasing the island's water retention and its arable land. Depicting the desperate situation, a Library of Congress study says: "Because of its lack of mineral wealth or petroleum resources . . . Haiti is forced to look to outside sources, principally the United States, to help to halt the environmental degradation which may soon lead to complete ecological—and human—disaster."[24]

Israel's first regional Environmental Protection Agency was formed at Haifa to cope with that port city's intolerable pollution. The young nation's pell-mell industrial growth ignored environmental safeguards until the stench and sludge could no longer be overlooked. Polluted with chemical wastes

22

from factories, the Kishon River is a prime target for rehabilitation.[25]

Pollution invariably accompanies massive industrialization. Tokyo police frequently carry oxygen and gas masks for their duties in fume-wreathed traffic. Mexico City has one of the highest air pollution levels in the world. "Acid rains" fall on Scandinavia, apparently wafted from England's tall smokestacks that propel sulfur dioxide into the sky where it mixes with water to form sulfuric acid. Priceless painting and statuary in Venice, Italy, deteriorate under the steady assault of polluted air. Similar acid-rain effects have been documented in the Great Lakes and the northeastern states of the U.S. where prevailing westerly winds dump pollutants picked up over the industrial cities. Some scientists refer to "aerial sewers" that mark the northeastern states as the discharge point.

Before his death in 1974, aviation pioneer Charles A. Lindbergh shrank from the rampant depredations of technology which he helped to stimulate. A late-blooming conservationist, he wrote in his unfinished *Autobiography of Values*, "Although I could find no wise alternatives, each year that I worked on weapons development left me more concerned about our future . . . . Civil technology vied with military technology in breaking down human heredity and the natural environment. Every day, increasing numbers of bulldozers and trucks tore into mountains, slashed through forests, leaving far greater scars on the earth's surface than those created by bombs."

Shortly before his death, the famed pioneer wrote: "I do not want to be a member of the generation that through blindness and indifference destroys the quality of life on our planet."[26]

The ravages of unrestrained technology are also recognized by Lewis Mumford, noted social historian who in a comprehensive study traced the rise of modern cities. He declared: "Nothing less than a profound reorientation of our vaunted technological 'way of life' will save this planet from becoming a lifeless desert."[27]

Sobering words. Are they accurate? "Doomsayers" attach no date to their warnings; they simply and starkly say that our present path leads inevitably to destruction. Some concede that civilization can turn to a safe course if enough citizens perceive the danger and take remedial steps.

Many technocrats, meanwhile, remain irrepressibly optimistic. In 1976 Herman Kahn and two colleagues at the Hudson Institute reiterated predictions made earlier in the book *The Next 200 Years*. "In A.D. 2176, we expect a population of 15 billion, with gross world product at $300 trillion, and per-capita income at $20,000 in 1975 dollars."[28]

These prophets brushed aside such perplexities as population pressure, environmental damage, and energy shortage. Each of these social quandaries met a swift verbal solution from the latter-day prophets, but the problems continue to intensify around the world.

Kahn and Wilfred Beckerman, another advocate of rapid industrial growth, weight their scale of social values on the side of technological and economic advances rather than on the side of human health and preservation of earth's life-systems. Kahn, a physicist by training, and Beckerman, a political economist, seem to see technology as society's savior.

Time proves that prophecy is a risky venture. Back in 1975 Beckerman wrote: "It is unlikely that the oil price will stay at anything like its early 1974 level for very long"—implying a dip back toward 1972 levels.[29] Four years later, the oil price was far higher. Similarly, Kahn assured the world in 1976: "Except for temporary fluctuations caused by bad luck or poor management, the world need not worry about energy shortages or costs in the future."[30] Those "temporary fluctuations" now look like constant escalation, and informed people around the world are justifiably concerned.

Donald Moffit, an editor of the *Wall Street Journal*, comments on the industrial growth controversy: "The ground for Kahn's optimism is not fact, but faith—his faith in progress . . . . But now, as even Kahn agrees, the world has reached a historic turning point . . . . If foremost among the governing ideas was the idea of progress, we cannot be so certain that it will continue to prevail."[31]

The noted British scientist and novelist C. P. Snow was asked the main difference between the world he grew up in and today's world, and he replied, "The absence of a future."

The specter of nuclear annihilation was Snow's great dread, but now, in addition, we face the possibility of mass poisoning by technological pollution.

Dire as these threats are, the greater menace, in the opinion of the authors of this book, was cited by Albert Einstein when he said, "We live in a time of perfected means and confused ends."

The late British economist, Ernest Schumacher, elaborated on this theme. "Science and engineering produce 'know-how,' but know-how is nothing by itself; it is a means without an end, a mere potentiality, an unfinished sentence. Know-how is no more a culture than a piano is music."[32]

Technicians and scientists have proved their productive ability, but leaders of public opinion and political policy have not formulated goals that challenge humanity's best efforts. While scientists open new material frontiers, cultural leaders must elevate human values. To maintain a civilization worth celebrating, ethically informed citizens must shape the goals of society. Scientific inventiveness must be guided by humanitarian principles.

Who should decide the complex questions on environmental regulations, economic expansion, and human well-being? In a pluralistic society many value systems contribute to final choices, but wise choices will be based upon proven principles. Adherence to a working set of principles will permit progress to be measured as we manage global resources in ways that improve the quality of life for all people.

The authors of this book believe that God created the earth for mankind to manage and enjoy, and that the essential principles for managing the global enterprise are given in the Judeo-Christian Scriptures. Ignorance and disregard of these principles have dangerously threatened human survival.

Wise management of Earth's resources will enable us to use technology to benefit all people within a healthy environment. This book elaborates this thesis and marks some practical pathways for action. But first we must comprehend the severity of Earth's ecological illness.

TWO
HEALTH
IN THE
CHEMICAL
CAULDRON

Is the global village seriously threatened by pollution, over-population, and energy shortage? Could the furor over energy shortages, unsafe drinking water, and radioactive wastes reflect the inordinate fears of alarmists or the mania of publicity seekers? British physicist John Maddox is one expert who derides the pessimistic prophets.

"Although these prophecies are founded in science, they are at best pseudo-science," Maddox alleges. "Their most common error is to suppose that the worst will always happen. And, to the extent that they are based on assumptions as to how people will behave, they ignore the ways in which social institutions and humane aspirations can conspire to solve the most daunting problems."[1]

But if calamity prophets are pessimistic in temperament, Maddox appears to be a congenital optimist. For while he admits the existence of severe environmental threats, he predicts future solutions on the basis of past successes without acknowledging that the warnings and actions of alarmed citizens were crucial factors in overcoming the dangers to society.

Confirmed optimists may look at past progress and chorus: "We've come a long way, baby." And pessimists may read the small print and rasp: "Invite the Surgeon General to my funeral." But realists will investigate today's scene and plan tomorrow's action accordingly. In this era of quantum changes, that is an extremely intricate task. Yet it is the indispensible requirement for preventing further breakdown in the links of a technologically strained environment.

The tiny island of Bahrain in the Persian Gulf is symbolic of today's deteriorating Earth. Long regarded by Arabs as the site of the biblical Garden of Eden, this spring-fed oasis of date palms, alfalfa, and pomegranate trees is drying up. Salty water from the sea, affected by the removal of oil from the island's wells, has invaded the bubbling springs and spread death to once-flourishing groves of date palms. The stench of oil and sewage hangs in the air instead of the fragrance of blossoms and sea spray.

Intensive oil production and rapid industrialization have replaced the island's gardens and gazelles with pipeline grids and sprawling factories. Bahrain's young generation is aware of the widening threat to its future. One seventeen-year-old student lamented: "Factories drain their wastes into seas and rivers. There will be no more pure water to drink, and it would need a lot of money to build more factories to purify the water. The wastes of factories are eaten by fish—then people eat those fish and get sick, too. The world might be a terrible place to live by the time I am forty."[2]

No scientist can accurately predict the health conditions in Bahrain or any other place on earth in twenty years. Some toxins reveal their deadly effects in humans only after the passage of many years, and the constant addition of new industrial compounds to the environment generate unpredictable consequences. Researchers have established, however, definite connections between certain pollutants and human disease.[3] Their discoveries of chemical threats to humans were aided by clues found in the destruction of simpler forms of life.

BREATHING WITHOUT MASKS

One of the earliest detection devices of air pollution was the lichen, a symbiotic association of an alga and a fungus plant. It is such a sensitive indicator that different lichens die at specific levels of pollution. Therefore, lichen populations became a useful monitor of air pollution.

Research on Long Island documented the adverse effects of New York City air pollution on lichens as far away as forty miles. Not only were some species smothered, but the vitality of surviving plants was reduced.[4] Similar results have been recorded in England and Wales.[5]

A striking instance of plant destruction from polluted air occurred in the Copper Basin at Copperhill, Tennessee. There, sulfur dioxide fumes from copper smelters created an eroded, lifeless desert extending over 7,000 acres. An additional 17,000 acres supports only sparse, grassy vegetation.[6]

Sulfur dioxide gases from auto exhausts and coal furnaces have reduced Illinois soybean yields by 5 percent. These results, obtained in 1977 and 1978 by Argonne National Laboratory scientists, could deter the expansion of coal-fueled industry.[7] Sulfur dioxide, changed into acid rain, intensifies soil acidity, prevents growth of some plants, and adversely affects water quality in the Great Lakes—the largest reservoir of fresh water in the world.

Many other pollutants are known to harm plant growth, and one of particular concern to urbanites is ozone. In the upper stratosphere, ozone forms a protective shield that filters out excessive rays of ultraviolet sunlight which causes mutations in living cells. The possible breakdown of this shield by jet contrails of supersonic airplanes and the fluorocarbon of aerosol spray cans influenced the U.S. government to cancel SST production and ban fluorocarbon propellants. But growing concentrations of ozone at ground level have become a "resource out of place."

Ozone multiplies in heavily motorized urban areas, impairing respiration of the inhabitants and damaging plant growth. Ozone in Los Angeles smog blighted Ponderosa pines twenty miles east of the city limits.[8] "Ozone alerts" in other large cities such as Chicago warn residents with respiratory or heart ailments to slow down until the pollution has dispersed, which often takes several days.

We have known for a long time that "bad air" can kill. Prior to sophisticated sensing equipment, a caged canary was the constant companion of underground coal miners. As the canary's song dwindled, miners became aware that lethal gases might be present. The incidence of black lung disease in coal miners is still a major issue in labor negotiations between unions and management.

Americans have been as reluctant to accept evidence that many health problems are environmentally caused as they were to heed the U.S. Surgeon General's warning on cigarette packages. Evidence in animal tests as well as the comparative incidence of lung cancer in smokers and non-smokers established the causal link. The case for environmentally caused cancer has been impressively given by health scientist Samuel Epstein in *The Politics of Cancer*. "An informed consensus has gradually developed that most cancer is environmental in origin and is therefore preventable."[9]

In the early 1970s several animals at the Staten Island Zoo in New York became ill and died. No evidence of any known disease could be discerned, but high levels of lead were found in blood samples, hair, and feces of the dead animals. Ralph Strebel, pathologist at New York Medical College, concluded, "Most of the lead taken in by the animals resulted from atmospheric fallout."[10] Lead in Greenland's icecap doubled between 1930 and 1965, probably from the burning of fossil fuels in distant countries.

Such results spurred environmental specialists and government agencies to push auto manufacturers to develop engines that would use lead-free gasoline. Yet the number of autos still using leaded gas steadily adds to the buildup of lead in the environment.

Large quantities of tetraethyl and tetramethyl lead—components of "regular" gasoline—have the following toxic effects on humans, according to health scientist George Walbott: "Patients exposed to these compounds even for short periods show nervous irritability, sleeplessness, terrifying dreams, emotional instability, and gastrointestinal symptoms with vomiting and diarrhea. With more intensive exposure, the victims of the disease become irrational and develop delusions and hallucinations with dramatic suddenness."[11]

Of all the careful studies reporting on the health effects of pollution, one of the most definitive appeared in *Science* magazine in 1970.[12] In this thorough survey of published research, the conclusions of Lester Lave and Eugene Seskin were cautiously worded but unequivocal.

For a number of countries, researchers found that air pollution "accounts for a doubling of the bronchitis mortality rate for urban, as compared to rural, areas." In a comparative study of St. Louis, Missouri, and Winnipeg, Canada, the incidence and severity of emphysema for those over twenty-five years of age was much higher in St. Louis, the city with higher levels of air pollution (e.g., in the forty-five-year-old group only 5 percent of those in Winnipeg but 46 percent of those in St. Louis showed evidence of this lung disease). One of Lave and Seskin's startling claims was: "Approximately 25 percent of mortality from lung cancer can be saved by a 50 percent reduction in air pollution, according to the studies cited."

Many environmental scientists believe the annual fall ritual of leaf burning adds dangerous pollutants to the air. Ron Musselman, assistant director of the Environmental Health Resource Center at the University of Illinois Medical Center, asserted in an interview that leaf burning produces benzopyrene, a potent carcinogen having the capability of inducing cancer in a great variety of tissues.

WATER UNFIT TO DRINK

Unless pollutants are kept from the atmosphere, many of them will end up in the lakes and rivers that provide water supplies for some cities. The Mississippi River is a notorious example. More than sixty industries pour all types of synthetic wastes into the Mississippi upstream from New Orleans. In 1976, Science magazine noted the high correlations of cancer mortality rates in Louisiana with the impurity of the drinking water obtained from the Mississippi River.[13] The analysis held not only for overall cancer incidence but also for cancers of the urinary and gastrointestinal tract. The chemical cauldron of today's industrialized society is obviously bubbling out of control.

A century ago, bacteria causing typhoid and other diseases were readily traced to sewage-contaminated water supplies, but the variety of potentially harmful chemicals that enter the nation's waterways today makes it difficult to isolate causal agents. Many that have not been shown to be human killers are proven destroyers of other forms of life.

Chemical pesticides such as DDT (dichlorodiphenyltrichloroethane) and PCB (polychlorinatedbiphenyl) have harmful effects in the lower stages of the food chain, but the potency heightens as the concentration increases in the higher levels of fish, birds, and humans. Sensitivity to the toxicity varies among different species, but both of these chlorinated hydrocarbons adversely affect reproductive systems. More seriously, one study revealed that workers in a DDT factory had a high incidence of liver and cardiovascular disease as well as neurologic malfunctions.[14]

The evidence for toxic effects from PCBs in Japan is so convincing that "Yusho disease" has been named for the symptoms. According to G. L. Waldbott, "The disease may have after-effects such as permanent disturbance of the central nervous system—changes in the heart and blood vessels and deformities of fingers, toes, ankles, wrists, and vertebrae, accompanied by pain, which may linger on for long periods of time." By the time people came down with the illness, 700,000 chickens had been killed by PCBs.[15]

In the United States, PCBs were added to adhesives, plastics, paints, and other surface coatings until Monsanto Industrial Chemicals Company began to sharply restrict PCB use in 1971. Unfortunately, the Outboard Marine Corporation in Waukegan, Illinois, had already discharged so much PCB into its harbor that the harbor floor was heavily contaminated to a depth of six feet. The federal Environmental Protection Agency (EPA) took Outboard Marine to court seeking a ruling as to who will pay an estimated $5 million for dredging and disposal costs in the harbor.

Heavily industrialized Japan suffered the first outbreak of methyl mercury poisoning. Birds and cats were the early victims of the debilitating affliction, then humans began to lose body coordination, go into convulsions, and die. The strange blight was centered near Minamata, a seacoast town depending on fishing and chemical manufacturing for jobs. Over a period of years, the cause was traced to the chemical industries that were discharging large quantities of mercury waste into the sea. Fish absorbed the mercury, and humans with high fish diets accumulated the poison to destructive levels. By 1970 (nearly twenty years after the first appearance of degenerating symptoms) over one hundred people suffered severe cases of "Minamata disease" and forty-one of them had died.[16]

Finally the Japanese government accepted evidence that the chemical industries were responsible for poisoning the mercury victims and began financial compensation. By 1975, health surveys located over 3,000 sufferers of the disease. Mercury, like many other chemicals used in industrial and agricultural operations, is widely distributed in nature, though at low concentrations. Through tragedy, we have learned it must be handled with care and its levels carefully monitored in water supplies and in fish sold for human consumption.

A distressing case of continuing water pollution brought notoriety to Lake Superior. Asbestos-like fibers from iron ore residues slide regularly into Silver Bay from operations of the Reserve Mining Company. Lengthy court proceedings and 18,000 pages of testimony failed to gain a judgment against the company in 1974, though it was known that asbestos fibers cause a form of cancer in the lungs and gastrointestinal tract. EPA scientists found fibers in Duluth's drinking water that are similar to those in Reserve Mining's residues, yet the threat to health continued during numerous appeals of court decisions.[17]

The latent period for the appearance of this mesothelioma cancer is twenty to thirty years, but before that time Reserve Mining may have departed the scene after depleting the area of ore.[18]

New Jersey, one of the nation's most industrialized states, also is one of the most cancer-ridden. Its cancer-death rate for men was 17 percent above the national average during the years 1950-68. In Salem County, where chemical plants crowd one another along the Delaware River, the mortality from bladder cancer for white males is the highest in the United States.[19] Known carcinogens are among the hundreds of chemical combinations discharged in dumps and water recharge areas, a practice the government is striving to eliminate.

No single agency polices the creation of new industrial compounds, and it is estimated that more than 1,000 additional synthetic substances are produced each year. Though regular testing of some substances such as food additives and medical products is conducted, countless chemicals enter our environment with no testing of their health effects. Geneticist James F. Crow declares: "There is reason to fear that some chemicals may constitute as important a risk as radiation, possibly a more serious one. . . . To consider only radiation hazards may be to ignore the submerged part of the iceberg."[20]

Some food additives, such as the nitrogen compounds added to processed meats to prevent bacterial contamination, are suspected cancer-causing agents. But the evidence gained in animal tests is often inconclusive, and the chemical may be present in greater quantities in the natural environment. Many factors must be judiciously weighed before specific effects are known.

THE SOIL IS DIRTY

Chemical pesticides were lauded as miracle workers by farmers several decades ago. Today they are recognized by environmentalists and farmers alike as a mixed blessing. They kill pests that destroy vegetable and fruit crops, but they often kill beneficial insects and birds that feed on the pests. In addition, some insecticides are very toxic and have long lives before decaying to stable form. DDT is one of these, as well as mercury compounds which have been a major component of fungicides used in seed preservation. Less-toxic pesticides are coming into wider use, and total amounts used may taper off as ecologists stimulate pest control by natural processes of the ecosystem.

Pollutants in the air and water often move to the soil and then into crops. When cadmium travels this route, the result can be very serious. Cadmium is refined from zinc and copper ore for industries that use electroplating processes, manufacture polyethylene pipe, or produce superphosphate fertilizer. Cadmium has been implicated in hypertension and kidney impairment in a number of animal studies.[21]

In Japan, the cadmium from a mining operation was carried

down the Jintsu River into the soil of rice paddies and subsequently taken into humans through the food chain. Symptoms of chronic poisoning appeared in impairment of kidney function and respiratory organs, and a softening of bone tissue through a disturbance of calcium metabolism. Soil chemist Geoffrey Leeper summarized the tragic consequences: "The most alarming evidence that is quoted about chronic cadmium poisoning is its appearance as the 'itai itai disease' on the Jintsu River in Japan, where about one hundred people have been crippled and have died prematurely."[22]

A variety of other substances, such as selenium, vanadium, and the chlorinated organic compound, carbon tetrachloride, have been associated with cancer and heart disease and generally poor health. Research is continuing along many fronts to ascertain the effects of these chemicals and many others upon living organisms.

Michigan suffered a chemical disaster in 1973 that may reverberate for years to come. Chemical analysts searched almost a year before identifying miniscule traces of virulent poison in livestock feed supplement that was curtailing milk production in some dairy animals and killing other farm animals. A flame-retarding chemical used in plastics manufacture, polybrominated biphenyl (PBB), had been mixed accidentally with a protein concentrate sold to an unknown number of farmers. Animal tests over the next five years forced the destruction of 35,000 cattle and a million pigs, chickens, and sheep infected with the deadly PBB.

Some farmers say even their replacement herds have become contaminated, perhaps by grazing in the areas where predecessors lived and died. An estimated $75 million worth of livestock has been lost to this single chemical blight, and human health has been seriously threatened, with the end not yet in sight.[23]

INVISIBLE POLLUTION: RADIATION

Another serious health hazard that must be prudently managed is man-made radiation. It increasingly penetrates all areas of the environment and in many instances will remain a deadly threat for generations to come.

Natural radiation has occurred since the formation of radioactive rocks at the origin of the earth. However, the nature of the process and its discovery in the element radium is generally credited to Henri Becquerel and Madame Curie about 1900. Since that time both natural and man-made radioactive elements have been used in a wide variety of applications ranging from therapeutic medicine to the dating of moon rocks brought back to earth by astronauts.

Portents of hazard from radioactive substances were scientifically confirmed by Jorma Miettinen, a Finnish ecologist, during the second national radioecology conference at the University of Michigan in 1967. Fission products from atmospheric testing of nuclear devices, in this case strontium 90 and cesium-137, had entered the lichen-reindeer-human food chain in Lapland. The World Health Organization collected data on the amount of radiation in the affected human population, revealing a high incidence of leukemia, blood diseases, and bone cancer in this rural society isolated from modern civilization.

The human toll from atomic bomb explosions at Hiroshima and Nagasaki in 1945 is still surfacing. While incidence of leukemia has dropped off, solid tumors such as bone cancer are now increasing in people who lived in the blast areas.

Flying high in the stratosphere today are numerous man-made satellites containing radioactive components. They are engineered to stay in orbit five hundred to one thousand years, but accidents do happen. The Russian Cosmos Satellite 954 that plummeted to Earth in early 1978 sprayed radioactive fragments over a wide area of northern Canada. Fortunately, few humans were near the debris, but Canada amassed a $7 million cleanup bill to send to Russia.

Cosmos 954 was one of an estimated dozen uranium-fueled satellites put into orbit by Russia. The U.S. has over half a dozen plutonium-powered satellites in the skies for navigation and communications duties. Supposedly these radioactive power plants can be disintegrated in space to prevent their unscheduled return to Earth, but the safeguards that failed on Cosmos 954 have yet to prove their effectiveness. Almost four decades after man's first sustained chain reaction of atomic fission, nuclear energy remains one of the greatest hazards to a healthful environment.

As nations develop and stockpile nuclear armaments, the construction of nuclear power plants mocks the hopes of those who campaigned for "the *peaceful* uses of atomic energy." Resistance by apprehensive citizens in both the U.S. and abroad has moved into the streets. In Europe popular protests have been gaining momentum, and construction of nuclear power plants has slowed drastically in West Germany. In May of 1979 global coordination of anti-nuclear groups resulted in the largest participation of protestors to date, estimated at over one million.

In the U.S., economic as well as environmental and health factors have put a crimp in plans for building 200 reactors by the end of the century. Construction time from initial plans to on-line power may take up to twelve years because of environmental impact assessments and necessary safety requirements. The 1979 accident at the Three Mile Island nuclear plant confirmed many suspicions and increased the reservations of many still uncertain about nuclear power. Costly delays and inflating prices make other power alternatives such as solar, coal-burning plants,and bio-gas generation from vegetation and organic wastes more feasible.

POLICING POLLUTION

Scientific studies are usually cautious about linking environmental pollutants with specific health problems. In some cases studies are tentative because of political or economic considerations, but in others such reticence reflects the complex nature of scientific investigations which make unequivocal conclusions unlikely.

Rather than wait for incontrovertible evidence, however, many conscientious citizens should promote reasonable measures that reduce suspected pollutants and should monitor the incidence of health problems in the affected areas. Tragedy may occur if environmental decisions are left to bureaucratic agencies, industries that place profits ahead of public health, or self-serving professionals who resist controls by saying: "That's not been proven yet." Aristotle, the ancient Greek philosopher, speaks to today's emergency: "The ultimate end . . . is not knowledge, but action. To be half-right on time may be more important than to obtain the whole truth too late."

Regulation of polluters by legislation and surveillance of the environment by citizen committees are crucial. Yet the key to restoring our damaged environment and our health is radically new strategy and tactics. We must recognize recyclable pollutants as resources out of place and consequently restore them to Earth's life-systems. This will prod ecological renewal and simultaneously spur economic activity in both rural and urban areas. Steps toward these goals are explained in the following chapters—steps that are being implemented on an ever-increasing scale each day.

THREE
NATURE'S
COMMUNITY

The air you just breathed is a basic natural resource. It is owned by no government, corporation, or individual, and so it is yours to use freely—as long as it is available and fit for human use. And there's the catch—clean air is dwindling because modern industry spews great quantities of noxious fumes and particles into the air which irritate, nauseate, or incapacitate humans in heavy concentrations. Since clean air is essential to health and belongs to everyone, despoilers of the atmosphere should answer to the public that is threatened by pollution.

Planet Earth boasts a marvelous clean-air system. Green plants both on land and in water take in carbon dioxide from the air and give off oxygen (mostly during daylight hours). Humans and animals, in turn, do the opposite—they inhale oxygen and exhale carbon dioxide. During this exchange, plants manufacture carbohydrates, proteins, and fats, while humans and animals process these into energy reserves and body-building materials. Green plants are the food factories of the earth, and humans are their totally dependent customers.

Energized by sunlight in the presence of chlorophyll, plant cells change carbon dioxide from the air and water from the soil into carbohydrates, or sugars. Some of the carbohydrates are then changed into protein, fat, and vitamins by the addition of nitrogen, phosphorus, potassium, and a number of

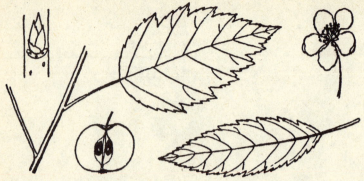

Earth's Food Factory: Green leaves like this crab-apple variety . . .

micronutrients from the soil. Humans and animals eat the nutrient-rich plants and, using oxygen, change the components into cell-building materials or use them for energy. Thus, solar-generated photosynthesis and oxygen-fed respiration activate the food-energy cycle of life.

Both food production in plants and food processing in humans are hindered by polluted air. The EPA estimates national crop losses from air pollution at $500 million each year.[1] Some airborne pollution particles settle in human lungs, aggravating respiratory ailments such as asthma and bronchitis. Other atmospheric pollutants have been identified in laboratory tests as possible causes of cancer and precursors to pneumonia.

Water is another basic natural resource. It is less accessible than oxygen and has been contaminated far worse than the air, so clean water has become a costly product in modern society.

In nature, water has its own cleansing mechanism. Rainwater seeps through the soil to groundwater reservoirs, leaving organic matter, viruses, and trace metals caught in the soil filter. Water in flowing streams is purified by microorganisms that change organic wastes to inorganic nutrients such as phosphates and nitrates—food for algae, which are food for fish, which are food for humans.

The natural cycle worked well until growing cities and multiplying industries overloaded waterways with human and factory wastes. As nutrient-stimulated algae proliferated, oxygen-starved fish and bacteria died, and odorous scum replaced clean water.

Before cities started using rivers as sewers, fishermen could see to the bottom. History recounts Captain John Smith scooping out shad with a frying pan at Great Falls on the Potomac River upstream from our nation's capital. Today a skillet would come up grimy.

Water purification and sewage treatment plants have been built in most cities to reduce pollution, but as growing cities channel more wastes to the waterways and industries discharge new chemical concoctions, these plants have proven inadequate. Thus clean rivers are a memory of the past, and the probability increases that humans will drink lethal or disease-inducing toxins in their water supplies. Clean water is essential to health since it is the vehicle that circulates nutrients through all life forms.

Nutrient-rich soil is the third resource bank that supports life. Though few of us need to be full-time farmers in the industrial era, the world needs increasing amounts of food from fertile land to feed its swelling population. Growing crops require nitrogen, phosphates, potassium, and many micronutrients from the soil.

Nitrogen makes up 79 percent of the atmosphere, but it must enter plants through the roots in nitrate form in order to join carbon, hydrogen, oxygen, and other elements in the formation of proteins, the building blocks of life. Soil bacteria are the primary agents that change most nitrogen into nitrates. These bacteria are especially productive when symbiotically incorporated into the roots of legume plants such as peas, beans, alfalfa, and clover. Before the widespread use of manufactured fertilizers, legumes were the farmers' "swing crops" to periodically replenish nitrogen in the soil. For rice paddies in the Far East, blue-green algae perform the nitrogen-to-nitrate step.

Phosphates come from sedimentary rocks, dead organisms, and animal wastes that are decomposed chemically by soil bacteria. Phosphorus, potassium, sulfur, calcium, magnesium, and other chemicals move through the food chain of plants, animals, and humans and then are returned to the soil by these useful decomposers of organic matter. Small amounts of metals such as zinc and copper which are necessary to human metabolism also travel this cycle.

Soil that lacks these mineral nutrients and bacterial organisms must be enriched by animal wastes, organic humus, or manufactured fertilizers. In recent years the development of intensive commercial fertilizing has dramatically increased food production, but is has simultaneously polluted aquatic ecosystems and consumed vast amounts of energy in fertilizer production. Unlike organic materials, manufactured fertilizers cannot build the loose, aerated topsoil and humus needed by plants to retain oxygen and moisture.

Heavy application of commercial fertilizers pollutes watercourses through rainwater runoff. Algae proliferate on the excess nitrates and phosphates, and their abundant growth soon cuts out light to plants below the surface. The dead plant material then utilizes all the available oxygen in the decomposition process, and remaining animal and plant life soon die. This type of pollution is a case of misplaced nutrients or resources, not toxic intrusion, yet it unbalances the local ecosystem and in many instances creates a cesspool with inadequate oxygen to sustain aquatic life.

COOPERATING WITH NATURE

Environmental care has been employed throughout history in two ways: dispersal of pollutants away from the originating point; and recycling of pollutants into the ecosystems that can change them back into resources. Many toxic pollutants of the technological era cannot be readily recycled, and their benefits to society must be scientifically measured against their harmfulness to life. When possible, energy-laden pollutants must be rechanneled for safe use. Modern environmental management is based on three principles that rule the natural lifecycles of the air, the water, and the land.

The first principle is that air, water, and land are components of a single, interacting environmental system. Simply put, everything is related to everything else.

The second principle is that the environmental system is virtually closed: nothing escapes, though it may change form. Everything has to be some place, and we need to determine where and in what form the unwanted things—wastes, for example—are going to be. This principle helps us ask the proper questions regarding technology and other human activities that influence environmental quality. It also helps us avoid programs—sometimes mislabeled "solutions"—which merely relocate the problem or alter the nature of the problem, such as taking pollutants out of the air and putting them into waterways.

The third principle is that many pollutants are potential resources out of place. Consequently, pollutants are to be identified, their economic and social values assessed, and their healthful disposition, abatement, or elimination charted.

Cooperating with nature, environmental management capitalizes on efficient recycling of resources and seeks substitutes for non-renewable resources which pollute the environment. In our technological age, substitution of renewable resources means the capture of solar energy for heating and cooling, the funneling of wind power into electrical generators, the addition of biogas power plants to stock farms, the application of organic sludge to farmland, the return of nutrient-rich waste-water to crop lands, and the preservation of flood plains and "green belts" near urban areas.

Environmental management also calls for refined technological processes: more efficient auto engines, adoption of water-saving plumbing devices, recycling of solid wastes, and diminishing dependence on radiation-producing nuclear power plants. Barry Commoner's recent book, *The Politics of Energy*, insists the energy crisis can be resolved only by switching from nonrenewable to renewable sources of power.[2]

TURN ON THE SUN

Solar energy is the great unharvested power for the future. Its supply is unlimited, it is environmentally clean, and it is increasingly competitive with fossil fuel costs.

Approximately 20 percent of the energy used in the United States heats and cools residences and business establishments. Technology now exists for solar energy to meet 50 percent of this demand, offering the opportunity to reduce overall U.S. consumption of polluting fossil fuels by 10 per-

Solar-heated House: The new look in sun-heated homes helps buyers pay off the mortgage.

cent. This would save one billion barrels of oil per year. At the same time, switching to solar energy systems would rejuvenate the national economy by enlisting countless engineers and technicians for designing equipment, laborers for manufacturing it, and various kinds of tradesmen for installing it. At present rates of fossil fuel cost (which are certain to increase), purchasers of solar equipment will recover their high initial outlay in from five to ten years through savings in fuel expenses and tax credits. Solar heating is a sound investment for builders and homeowners in our high-priced energy era. (See chapters 5 and 6 for specific examples.)

Future technology is expected to convert solar energy directly into electricity for powering factories, but that prospect cannot meet current industrial needs. Today significant savings in energy can be attained through energy conservation, the use of alternate energy sources, and the application of present solar technology. Revolutionary solutions to our massive energy problems may even now be germinating in one or more laboratories of the scientific community.

WASTE TO WEALTH

The second major alleviation of the pollution-energy crisis can be realized through recycling wastewater. The nation-wide practice of pumping sewage into waterways is probably our most harmful and costly environmental crime for these reasons: (1) nitrate-and phosphate-laden sewage smothers the life-bearing food chains of aquatic ecosystems, (2) discarded sewage is rich with nutrients needed by land plants, (3) nitrate-starved farmlands have to be replenished with high-priced, oil-based fertilizers which promptly end up polluting streams, and (4) tests prove the land-treatment system of sewage disposal is the most economical and healthful method of producing clean water.

Most urbanites don't realize that their grandfathers—if they were farmers—used animal manure to fertilize crops, and that farmers down through the ages did the same. Today treated sewage and sludges applied to the land are producing bumper harvests in many locations with no ill effects.

We have been conditioned to think of sewage as worthless and dangerous to health. This is partly true because some diseases are transmitted through human waste. But most of us never learned that soil is a super-filter for disease bacteria and that sewage contains elements essential for plant growth. Discarding these nutrients into rivers rather than returning them to the land is an unscientific travesty against nature and energy efficiency.

Scores of European cities operated sewage farms in the nineteenth century. In 1857 the British Royal Commission on Sewage Disposal declared: "The right way to dispose of town sewage is to apply it to the land . . . it is only by such application that the pollution of rivers can be avoided."[3] Melbourne, Australia, has enriched thousands of acres of pastureland for beef cattle and sheep since the 1890s with raw sewage, causing no adverse effects.[4]

Recycling treated wastewater from cities onto farmland would promote an environmental-energy revolution by simultaneously reducing energy spent in treating sewage, curtailing pollution, increasing food yields, and renovating the water better than can any chemical treatment plant. This carefully engineered system has been thoroughly tested in Muskegon

County, Michigan—among other places—and is ready for nationwide adoption.

In 1977 the U.S. Environmental Protection Agency and its administrator, Douglas M. Costle, recognized the primacy of water recycling and land treatment for meeting national clean water goals. In a memorandum to regional administrators and assistants Costle directed:

"Pursuant to President Carter's environmental message to the Congress, EPA must press vigorously for publicly owned treatment works to recycle and/or reclaim wastewater through the Construction Grants Program of the Federal Water Pollution Control Act Amendments of 1972 . . . It is apparent that the technology for planning, designing, constructing, and operating land treatment facilities is fully adequate to meet the 1985 goal of P.L. 92-500."

That goal is the cleanup of all U.S. waterways by 1985. To encourage the recycling of pollutants, Congress passed the Clean Water Act Amendments of 1977 that allocates 10 percent extra funding to communities that adopt the land treatment system.

Nearly 1,000 American communities currently fertilize crops with pretreated wastewater. Many of them are located

This cotton crop near Lubbock is being irrigated and fertilized with processed wastewater.

in the arid southwest where scarce water is apportioned by law. Bakersfield, California, and Lubbock, Texas, are two medium-sized communities that use sewage to irrigate crops. Yet most urban areas still consider rivers and oceans as part of the sewage treatment system rather than fragile links in Earth's food cycle.

Pollution from conventional treatment of sewage is recognized as a problem by ecologists, conservationists, tax-conscious consumers, and the U.S. government. With industrial and municipal wastes increasing in quantity and complexity, Congress has tightened pollution standards to protect health and improve environmental quality.

New requirements will force many cities to update sewage treatment facilities to meet clean water standards. They can either expand conventional systems that process sewage chemically and biologically then discharge it into waterways, or they can inaugurate the land treatment system that uses plants and soil as purifying filters and restores nutrients to the food cycle. The flush-sewage-down-the-river method invariably degrades the water and constantly requires more chemicals. The three-stage sewage plant consumes twice the amount of chemicals and energy as secondary treatment plants; projected costs for upgrading the nation's conventional sewage system is an estimated $100 billion in ten years.

Muskegon County in southwestern Michigan chose the land treatment method in the early 1970s. Area unemployment was high and morale was low in 1968. Though a lakeside community, the shoreland was scarred with abandoned sand-mining operations, the area's aging industries were struggling to compete against more modern facilities, and deforested lands had lost much of their topsoil to rivers and lakes. Surrounded by economic problems and facing a government order to clean up their sewage discharge, civic and industrial leaders united in a far-sighted project.

The County Board of Commissioners authorized a comprehensive plan for countywide management of wastewater by land treatment. Completed in 1973, the system pipes all municipal and industrial wastewater to lagoons in a large, open area where aerator pumps stir the water and bacterial colonies break down the organic material. Huge storage basins then receive the effluent, where bacteria continue their attack on pollutants. Chlorine is added when tests indicate the need to kill disease-carrying bacteria.

During the crop-growing season, the nutrient-rich water is pumped to irrigation rigs and sprayed onto crops. Filtering through the earth, the water loses metals and bacteria to soil particles, and underground drain pipes collect the excess water—now filtered to drinking water quality—and carry it away for re-use. The whole operation is carefully monitored and regulated for optimum results.

This system not only eliminated Muskegon County's sewage pollution and met government standards, it returned $300,000 in crop earnings from the irrigated land the first year. With the water-nutrient cycle restored in the Muskegon project, rivers and lakes are resuming their natural beauty and healthfulness for a delighted populace. Residents were even more gratified in 1978 by a $800,000 corn crop on their sewage-treated land. (See Chapter 6 for further description.)

Another benefit of the land treatment process was described by Benjamin J. Reynolds, a dairy farmer in Pennsylvania who witnessed improvements brought by the process to his area.

> By utilizing agricultural grounds for the disposal of treated municipal effluent and agricultural waste, we can create the greenbelts that our planners are dreaming of. This will provide areas of lush green crops and verdant woods which will be aesthetically pleasing to the eyes. These lush green fields and woods will provide open space for our ever-growing populations, increase food production, provide more nursery trees to beautify our landscape, provide more board feet of lumber for our use. As I like to term it, this would be tax-paying open space producing highly salable crops for our economy. I feel the strongest benefit besides the production of food and fiber will be the cooperation engendered between our urban neighbors and our rural communities.[5]

Some planners have feared that the national drive for clean water would bankrupt the nation. That seems entirely possible if municipalities rely on costly conventional sewage plants. But if clean water is achieved through multipurpose land treatment projects that consolidate industrial cooling, provision of recreational space, and production of agricultural and silvicultural products by recycling wastewater, then the nation can afford clean water.

A cost study of wastewater treatment systems for the Chicago metropolitan area is revealing. Based on actual bids and construction experience for (1) a conventional treatment plant (primary and secondary), (2) the same system with advanced (tertiary) treatment, and (3) a land treatment system, the third was found less costly for both capital outlay and operating expenses. In fact, the operating cost for land treatment was only one-third as high as for tertiary treatment.[6]

Wise environmental management will also turn sludge, the residue of sewage treatment, into profit instead of pollution. Conventional sewage plants collect sludge and dump it in a landfill or burn it. Incineration adds to atmospheric pollution, is very expensive in fuel costs, and squanders valuable nutrients.

To incinerate a ton of dry solids in sludge requires the natural gas equivalent of 108 gallons of #3 fuel oil. The process consumes supply-short fuel and spews particulate matter, hydrocarbons, and volatilized metals into the air, dispersing nutrients and sediment that could enrich and condition the ground. If incineration is adopted as the main method of sludge disposal throughout the country, nearly two billion gallons of crude oil a year will be used to burn it up (based on an estimated one ton of sludge per million gallons of wastewater and a daily flow of 40 billion gallons).

The Metropolitan Sanitary District of Chicago found a better way. In 1970 the District initiated a project to pump liquid sludge from lagoons into special tank cars that moved 800,000 gallons in a 40-car train to private agricultural lands 160 miles away. There the sludge was plowed into the ground as a soil conditioner and fertilizer that increased yields above that produced by commercial fertilizers.[7] Building upon this experience, the District developed other programs to transport sludge on river barges and trains to reclamation sites as far as 300 miles downstate.

49

Not all kinds of sludge are equally useful for this purpose. Some types of industrial waste contain toxic heavy metals such as cadmium which enter the human food chain through uptake from the soil, especially in leafy vegetables such as lettuce and swiss chard. Proper management of heavy metals in agricultural soils is explained by Geoffrey Leeper in his pioneering book, *Managing the Heavy Metals on the Land.*[8]

T.D. Hinesly, soil scientist at the University of Illinois, suggests that transport of liquid sludge by pipeline is another economically feasible method for large municipal facilities. He asks: "Why continue to squander our waste-resources when they can be forever productive? This is one of the *now* questions that should be answered without further delay."[9]

The obvious advantages of land treatment for sewage have been ignored or resisted by influential segments of our society. Sanitary engineers, manufacturers, and contractors associated with sewage disposal find it difficult to admit that old techniques cannot cope with new conditions. Concerned primarily with the plumbing phase of sewage sanitation, they have strenuously resisted change to land treatment methods.

In 1970 sanitary engineers proposed the addition of advanced waste treatment processes at the Blue Plains plant near Washington, D.C. to provide additional capacity and to mitigate the pollution in the befouled Potomac River. Some environmentalists countered with an appeal to consider land treatment of the sewage, pointing out that chemical tertiary treatment at its full capacity of 309 million gallons daily would consume nearly 2 million kilowatt hours of electricity, 500 tons of chemicals, and more than 100,000 gallons of fuel oil each day. In addition, they charged, the incineration of sludge residue would expose residents to health risks from airborne lead, mercury, particulates, and oxides of nitrogen and sulphur.

The head of the Virginia State Water Control Board denounced the dissenters in scurrilous terms. Subsequently the Interstate Commission on the Potomac River Basin published false allegations about land treatment in its newsletters of November-December 1971 and June 1973.[10]

But change is coming. In addition to the fiscal support of EPA for land treatment techniques, organized labor is picking up the option. George Roberts, consultant to the Colorado Labor Council of AFL-CIO, said, "Sewage treatment is the single biggest public works program in the country. We want to find out what the changes will mean to the construction industry. The alternative systems, like land treatment, will provide more jobs."[11] And these will be productive jobs, increasing national supplies of food and fiber, rather than jobs supported by community taxes that merely relocate pollutants.

Another source of unclaimed power is animal wastes which contain a prodigious amount of untapped energy, in addition to their plant nutrients. In biogas plants, the manure from only half of the nation's 130 million cattle and 55 million hogs could produce methane (natural) gas containing the energy equivalent of 100 million barrels of oil a year. Biogas plants are successfully operating near some large feedlots, and these small anaerobic digestors are producing all the electricity needed on some farms, pointing the way to massive savings of energy (see chapter 6).

BACK TO BASICS

Microbiologist Rene Dubos calls the deterioration of water and soil the prime factor in the decline of past civilizations. He acknowledges the havoc of civil strife, war, and disease in the collapse of cultures, but cites soil depletion and water exhaustion as the primary causes of their extinction.

Dubos points out that Japan and certain areas of Europe have cultivated their lands for a thousand years without bankrupting the soil, and their societies have endured despite wars and natural disasters. He observes: "The land has remained fertile under intense cultivation only where farmers have used it according to sound ecological principles. Unwise management of nature or of technology can destroy civilization in any climate and land, under any political system."[12]

Cooperation with nature is indispensible to human health, and stewardship of natural resources is critical in our era of food shortage, high fuel costs, rising economic expectancy, and accelerating technological change.

Alvin Toffler brands "avalanching change" as the most threatening aspect of contemporary culture. Expanding technology has propelled the United States and other developed countries from what he calls the first stage of economic development—agriculture, to the second stage—industrialism, and very recently to superindustrialism. With a majority of workers moving from food production to manufacturing and trades, and then to "white collar" supporting roles in society, Toffler declares that "the world's first service economy has been born."[13]

Though "service" instead of agriculture and manufacturing may be the main activity of modern workers, it is still true that "the profit of the earth is for all: the king himself shall be served by the field," as Israel's sage King Solomon affirmed (Ecclesiastes 5:9, KJV). As never before, food is a major concern of humanity, and clean water, air, and soil are necessary for abundant food production.

U Thant, former secretary general of the United Nations, noted the importance of intelligent resource management when he said, "The central stupendous truth about developed economies today is that they can have—in anything but the shortest run—the kind of scale of resources they decide to have. . . . It is no longer resources that limit decisions. It is the decision that makes the resources."[14]

This new reality makes environmental decisions crucial to the life and health of humanity. In these complex and urgent issues, what rationale or ethic should guide our decisions? Will it be protective policies by the privileged few against the deprived masses? Or quality-over-quantity measures achieved by pervasive government controls? Perhaps the wait-and-see philosophy of indecisive observers will prevail.

The authors of this book recommend the ethic woven throughout the Judeo-Christian Scriptures which give principles for harmonious relationships between humanity and deity, between man and man, and between humanity and the environment. Jesus Christ affirmed the interdependence of life forms and the universality of governing principles. He had pertinent words for physical life as well as spiritual life. He advocated a program of Earth management that is equally relevant for meeting all the needs of humanity or rescuing a civilization in jeopardy.

FOUR
THE HUMAN
COMMUNITY

"It's a big, wide, wonderful world . . ." wrote the lyricist. Countless marvels of the universe are obvious to everyone, but poetic eloquence and scientific insight have deepened our understanding and appreciation. Men such as William G. Pollard, an astute physicist and articulate clergyman, sharpen our focus with perceptions like these:

> I myself doubt that there is another place like the earth within a thousand light years of us. If so, the earth with its vistas of breathtaking beauty, its azure seas, beaches, mighty mountains, and soft blanket of forest and steppe is a veritable wonderland in the universe. It is a gem of rare and magic beauty hung in a trackless space filled with lethal radiations and accompanied in its journey by sister planets which are either viciously hot or dreadfully cold, arid, and lifeless chunks of raw rock. Earth is choice, precious, and sacred beyond all comparison or measure.[1]

Yet this wonderful world also has floods, deserts, rattlesnakes, and cockroaches. And we Earthlings sometimes spit out sand while savoring succulent shrimp. But planet Earth and its human inhabitants seem curiously well matched. For humans also are magnificent and repulsive. Often heroic in

53

danger, gallant in defeat, and selfless in suffering, mankind is perversely arrogant in success, unscrupulous in competition, and fickle in friendship. Humanity and the habitat share a similar majesty and malignancy. Their destiny seems closely intertwined.

If world population outstrips natural resources, say some environmentalists, widespread starvation is inevitable. The alternative they recommend is government control of population size. Other environmentalists warn that pollution will scourge humanity unless technology is drastically restricted. They may prove correct, but theologian Joseph Sittler makes a more hopeful analysis. It's a matter of choice, he points out.

"The problem of material is not a material problem, for man is in it, and he complicates every problem. The problem of enough to eat is not ultimately an economic problem. For as man confronts the marvelous richness of the earth he can use these riches or abuse them. Which of these he chooses is a matter not soluble by mere planning. For there will never be enough for both love and lust!"[2]

Man has the capacity for managing resources to provide food, clothes, shelter, and varied amenities to all humanity. Earth's prolific resources are beyond inventory; the question is: are Earth's managers equal to the challenge?

Alvin Toffler stresses the importance of comprehensive planning. He urges "a whole battery of compatible policies dealing not only with money supply, wages, prices, and balance of payments, but with everything from resource use and environment to education and cultural life, from transport and communications to the changing relationships between men and women."[3]

Professor Dennis Meadows proposes fundamental value changes in our established patterns. "Any deliberate attempt to reach a rational and enduring state of equilibrium by planned measures, rather than by chance or catastrophe, must ultimately be founded on a basic change of values and goals at individual, national, and world levels."[4] (See Chapters 7 and 8 for elaboration of this theme.)

In the opinion of Paul Ehrlich, visionaries must communicate their farsightedness to others if mankind is to win the life-or-death struggle.

Perhaps the major necessary ingredient that has been missing from a solution to the problems of both the United States and the rest of the world is a goal, a vision of the kind of Spaceship Earth that ought to be and the kind of crew that should man her. Society has always had its visionaries who talked of love, beauty, peace, and plenty. But somehow the "practical" men have always been there to praise smog as a sign of progress, to preach "just" wars, and to restrict love while giving hate free rein. It must be one of the greatest ironies of the history of the human species that the only salvation for the practical men now lies in what they think of as the dreams of idealists. The question now is: Can the self-proclaimed "realists" be persuaded to face reality in time?"[5]

Discovery of renewable energy sources, expansion of recycling techniques, and a radical sense of community can be our negotiable currency for meeting worldwide needs, in the view of University of Colorado economist Kenneth Boulding, president of the American Association for the Advancement of Science. Compassion would enable humanity to extend Earth's resources to more and more deprived peoples, according to Boulding. This would mean a supplanting of "lifeboat ethics" with a flexible "mesa" mentality, in which new sources of renewable energy would be shared with the world, gradually moving the sheltering "fence" of abundance toward the global cliff that imperils poor people until all humanity is safely encompassed.[6]

The late anthropologist Margaret Mead also called for a spirit of community that will unite segments of society in constructive efforts. "Unless we can use imagery that involves everybody in this country and ultimately everybody in the world, unless we can use both the notion of a spaceship and the notion of an island, both what man has made and what he has received as a precious heritage from nature, unless we can put these things together we will again fragment ourselves and again have people fighting with one another instead of working together toward a common goal— to protect the earth and its people."[7]

Wiser planning, greater effort, deeper compassion—these

all-out measures involve the minds, bodies, and hearts of humanity. Significantly, they speak of the whole man acting in concert with other men for the common good. They speak of intellect and bodies motivated by compassionate hearts. As historian Lynn White declared, the way out of our environmental dilemma is "essentially religious, whether we call it that or not."[8]

Religions, of course, are as diverse as their founders. How does Christianity, the dominant faith in the highly technological Western world, blend with ecological principles?

ECOLOGICAL ROOTS

If recognition of the value and interrelationships of flowers, birds, and people is the mark of an ecologist, Jesus Christ was an early model. He said the lilies are more lavishly adorned than King Solomon, and the common sparrow does not perish without the Creator's notice. As for man, his imperishable spirit makes him worth more than all the rest of creation (Matthew 6:28, 29; 10:29; 16:26).

By means of a story with multilevel meanings, Jesus taught the principle of efficient management of resources. In the illustration recorded in Matthew 25:13-29 a man in preparation for a journey to a far country delivered to his servants, "talents" or resources. He gave one five, another two, and another one. The one with five resources managed and used them and they grew to ten. The one with two resources put them to work and they grew to four. But the old-line conservationist in the group hid his resource. This one who did not productively manage his resource had it taken away by the owner and given to the one who had ten.

Jesus' scale of justice rewarded efficient management of resources. Personal diligence won additional opportunity and responsibility. The steward who merely "conserved" the resource from loss was condemned for his miserliness and fear. He added nothing to the legacy apportioned to him; even his time was wasted.

Although "conservation" may be defined as "the wise use of resources," it also has an unfortunate connotation of non-use, hoarding, or anxious preservation. The term "management," denoting the productive employment of resources, is a more positive and progressive concept. Management denotes action, organization, and direction.

All of us are managers of the rich resources of Earth. Some portion of these resources comes under our control throughout life, and we individually decide whether we will selfishly consume all we can accumulate or share with others in need.

Jesus showed his regard for the smallest of resources in a remarkable way. After a multitude of men, women, and children finished an instant meal of bread and fish served by Jesus, he directed his helpers to gather the leftovers so none would be wasted. We are not told by John (in John 6) how the leftovers were used, but Jesus' action implies a purpose and value for even the small things of life. It is humane—and divine—to respect each part of creation in its proper place.

As a Jew, Jesus lived by the Hebrew Scriptures. He frequently quoted Old Testament prophets and called attention to the laws of Jehovah. These laws and ordinances were amazingly comprehensive in relation to all facets of life.

The God of the Hebrews was no remote, incomprehensible deity. Through prophet-writers he communicated directives on almost every topic under the sun, including care of the earth. Consequently, the young nation of Israel migrating from Egypt to the land of Canaan that "flowed with milk and honey" was jeopardized more by spiritual unfaithfulness than by agricultural ignorance.

Israel's agricultural-religious regimen included replenishing the soil nutrients by plowing crop residues underground and letting the land "rest" every seventh or sabbatical year. This practice not only renewed the soil in a unique fashion, it turned the attention of the populace from prosperity supplied by fertile land to prosperity issuing from spiritual diligence throughout a ceremonial year.

The principle of recycling nutrients to the ground was reinforced by the requirement to locate toilet areas outside living areas and to bury human excrement (Deuteronomy 23:13). Not only was soil fertility improved but better human health was promoted. Inedible parts of animal carcasses and ashes from cooking fires also enriched the soil. The importance of sowing and reaping was recognized by everyone in the agrarian-pastoral economy, and careful management often made the difference between scrambling for necessities and enjoying the bounty of one's vines and fig trees.

"Adamah" is the Hebrew word for husbandry, or tending of growing things, indicating the essential vocation of Adam, the progenitor of mankind in the Hebrew Scriptures. According to Genesis 2:15, the Creator's original purpose for humanity included stewardship of the earth. "And the Lord God took the man, and put him into the garden of Eden to dress it and to keep it." The verb "dress," explain language experts, meant to serve the garden, not despoil it. Adam and Eve were appointed managers of their environment, accountable to their Maker.

"Be fruitful, and multiply, and replenish the earth, and subdue it; and have dominion over the fish of the sea, and over the fowl of the air, and over every living thing that moveth upon the earth" (Genesis 1:28). In this divine directive, humanity became the steward of the environment which either nurtures or nullifies life. Man's supervision was to extend as far as his ability and knowledge reached.

This biblical mandate has its critics. Lynn White has charged that the Judeo-Christian belief of man's superiority over nature inspired contemptuous exploitation of the environment.

"Both our present science and our present technology are so tinctured with orthodox Christian arrogance toward nature that no solution for ecologic crisis can be expected from them alone. Hence, we shall continue to have a worsening ecologic crisis until we reject the Christian axiom that nature has no reason for existence save to serve man."[9] White has obviously observed errant Christian behavior, but he has misinterpreted a scriptural imperative. The basic purpose of nature as revealed in Scripture is to glorify God, as the psalmist declared long ago (Psalm 19:1). That purpose is served as man and nature relate properly to each other.

As theologian Francis Schaeffer explains: "Man has dominion over nature, but he uses it wrongly. The Christian is called upon to exhibit this dominion, but exhibit it rightly: treating the thing as having value in itself, exercising dominion without being destructive. The church should always have taught and done this, but she generally failed to do so, and we need to confess our failure. Francis Bacon understood this, but by and large we must say that for a long, long time Christian teachers, including the best orthodox theologians, have shown a real poverty here."[10]

Professor White rightly deplores environmental depredations in the nominally Christian Western world. But greedy exploitation of nature is not limited to one part of the globe or one religious orientation, as Rene Dubos noted.

> All over the globe and at all times in the past, men have pillaged nature and disturbed the ecological equilibrium, usually out of ignorance, but also because they have always been more concerned with immediate advantages than with long-range goals. Moreover, they could not foresee that they were preparing for ecological disasters, nor did they have a real choice of alternatives. If men are more destructive now than they were in the past, it is because there are more of them and because they have at their command more powerful means of destruction, not because they have been influenced by the Bible. In fact, the Judeo-Christian peoples were probably the first to develop on a large scale a pervasive concern for land management and an ethic of nature.."[11]

Throughout history and across the world, ambitious men have over-cultivated the land, over-grazed pastures, and over-killed animal herds. Ancient Assyrian hunters exterminated an animal species merely for sport, and modern goatherders are enlarging the Sahara Desert by overgrazing in the Sahel region.

Currently a 300-square-mile area in northeastern Brazil is being ravaged by quick-profit entrepreneurs. German farmers moved from the mountains into a tropical forest near the ocean some years ago and cleared some of the land by burning trees. Lumber companies soon followed with chain saws, and hundreds of plant and bird species quickly disappeared. With the ecological system unbalanced, pests and parasites proliferated. "Within twenty years," reported environmentalist Augusto Ruschi, "the Atlantica forest was turned into pasture lands and coffee plantations, and now the area is marching toward desertification."[12] The recent discovery of large oil reserves deep in the Peruvian jungle portends another ecological assault in the rush to satisfy energy demands.

Such environmental wantonness sprinkles history in the East and in the West, within and outside the Judeo-Christian tradition. The process reminds Christians of the biblical principle: "Whatever a man sows, this he will also reap" (Galatians 6:7, NASB). Plant wheat and reap a wheat field; demonstrate kindness and gain a friend. Dispense pollution and suffer debility.

Adherents of Judeo-Christian ethics must admit that some of the "faithful" have been faithless in following their creed. In the past, "dominion over the earth" meant something like "might makes right" or "property is sacred," and many Christians avariciously collected the treasures of Earth while ostentatiously patronizing synagogues and churches. More authentic representatives of biblical faith, however, have enhanced the environment and shared their possessions in obedience to Jesus' command to "love your neighbor as yourself." Concern for neighbors of all kinds is the distinguishing trait of the genuine Christian community.

GOD AND THE GOOD EARTH

Proper understanding of the biblical mandate to exercise "dominion over the earth" is essential to beneficial relationships between man and his environment. Christian theologians and scientists have defined the relationship in various ways. The following representatives from history and contemporary life perceive a hierachy of relationships between the Creator, humanity, and nature that imparts value and purpose to all.

Augustine asserted: "There is a hierarchy of created realities, from earthly to heavenly, from visible to invisible, some being better than others, and the very reason of their inequality is to make possible an existence for them all. For, God is the kind of artist whose greatness in his masterpieces is not lessened in his minor works—which, of course, are not significant by reason of any sublimity in themselves, since they have

none, but only by reason of the wisdom of their Designer."[13]
John Calvin likewise emphasized God's wisdom in creation.

> Being placed in this most beautiful theater, let us not
> decline to take a pious delight in the clear and manifest
> works of God. . . . In order that we may apprehend with
> true faith what it is necessary to know concerning God,
> it is of importance to attend to the history of the creation,
> as briefly recorded by Moses, and afterward more co-
> piously illustrated by pious writers, more especially by
> Basil and Ambrose.
> From this history we learn that God, by the power of
> his Word and his Spirit, created the heavens and the
> earth out of nothing; that thereafter he produced things
> inanimate and animate of every kind, arranging an in-
> numerable variety of objects in admirable order, giving
> each kind its proper nature, office, place, and station.[14]

More recently, German preacher and author Erich Sauer
pointed out the necessity of knowing nature's laws in order to
manage Earth wisely. "Man should not be a tyrant over
nature. He should not misuse it by senselessly destroying
beautiful landscapes. We cannot rule over nature without
first discovering its laws. Man, conscious of his nobility,
should walk worthily of his high calling."[15]

Sounding a little like a Hebrew prophet, soil expert Walter
Lowdermilk added his warning to modern vandals of the good
earth. After pioneering soil conservation methods in the United States, China, and the Middle East, this American agrono-
mist suggested an "Eleventh Commandment" concerning man's
relationship to nature.

> Thou shalt inherit the Holy Earth as a faithful steward,
> conserving its productivity and resources from genera-
> tion to generation. Thou shalt safeguard thy fields from
> soil erosion, thy living waters from drying up, thy forest
> from desolation; and protect the hills from overgrazing
> by thy herds, that thy descendants may have abundance
> forever. If any shall fail in this good stewardship of the

earth, thy fruitful fields shall become sterile, stony ground, or wasting gullies, and thy descendants shall decrease and live in poverty, or perish from off the face of the earth."[16]

"Are the old agricultural laws of Jehovah a dead letter?" probes writer Alastair I. MacKay. He continues:

Has agricultural science and mechanized farming relegated into oblivion the ancient laws of the books of Moses? The billion-dollar chemical industry, aided and abetted by the powerful resources of orthodox agricultural science, keeps forcing production and variety at the expense of quality and vitality. The soil is regarded as a chemical test tube, insect and disease ravages a challenge to deadlier poisons (with a fingers-crossed attitude toward the health of the propaganda-dazed consumer), while mechanized farming and gardening beckons big business and elbows poor husbandman Adam out of his homemade garden of Eden.[17]

The religious root of our ecological crimes is uncovered by theologian Eric Rust:

Man has not fulfilled the divine image. His freedom and creativity have become misdirected by egotism and a lust for power. . . . With all his scientific knowledge and technological skill, man does not cooperate with the ecological balance of nature. He rapes nature, upsets its balance, exploits it for his own selfish ends. He becomes divorced from nature because he is divorced from God.[18]

EARTH'S MASTER-SERVANT

Christian theology teaches that all parts of creation have value, but different values and roles. Understanding the proper relationships between living entities is critical to their healthy development. Neither the anthropocentrism of so-called "exclusionist" theologians nor the pantheism-tinged naturalism of "inclusionists" represents biblical Christianity.

Frederick Elder analyzes these views in his book, *Crisis in Eden.*[19]

Humanity discovers its true role and destiny in the design of creation, contends theologian Carl E. Armerding.

> Up to the sixth day, with its creation of man, each natural element brought into being finds its meaning in fulfilling a role cast for it in the benevolent order of things. Light dispels darkness and we have day. The firmament keeps the waters separated. The dry land provides a platform for vegetation which in turn feeds all the living creatures. The seas become in their turn an environment for the fish and swarming creatures.
>
> The two great lights rule (or give order to) the principle parts of the cycle: day and night. And finally man, as the highest of the created order, serves to keep all of the rest in order, functioning smoothly. In fact, it is in Genesis 1 with its penchant for order and its transcendent and overarching concept of a purposeful universe, that a truly balanced cosmological system can be found. . . . Herein lies the origin of science and technology. . . .
>
> Man derives his meaning from God whose program, though it from the beginning offered man the kingdom, included a recognition of God's ultimate lordship over all creation and saw man as a responsible steward, not an independent tyrant. . . . There is no such thing for biblical man as unlimited freedom or unlimited rights. His freedom is that of the operator of a beautifully functioning machine. . . . But when he ignores the rules and decides he can ignore the complexities of his machine and the instructions left by its Maker, his freedom is lost and he becomes the destroyer both of the machine and his own function as its lord.[20]

Freedom of choice is one of the distinguishing qualities of man that separates him from other forms of life, points out Francis Schaeffer.

> Man's relationship is not basically downward but upward. Man is separated, as personal, from nature be-

63

cause he is made in the image of God. That is, he has personality and as such he is unique in the creation, but he is united to all other creatures as being created. Man is separated from everything else, but that does not mean that there is not also a proper relationship downward on the side of man's being created and finite. . . . This is a concept that no other philosophy has.

An essential part of a true philosophy is a correct understanding of the pattern and plan of creation as revealed by the God who made it. For instance, we must see that each step "higher"—the machine, the plant, the animal, and man—has the use of that which is lower than itself. . . .

I who am made in the image of God can make a choice. I am able to do things to nature that I should not do. So I am to put a *self*-limitation on what is possible. The horror and ugliness of modern man in his technology and in his individual life is that he does everything he can do, without limitation. Everything he *can* do he *does*. He kills the world, he kills mankind, and he kills himself . . . this is the problem all the way back to the Garden of Eden.[21]

Pantheistic religions of the East give a veneer of sublimity to all life by investing everything with divinity. By letting "sacred" cows eat their fill while people starve, Hindus fail to differentiate levels of value, thereby degrading humanity.

Intelligence enables humans to discern the proper ways to serve nature, fellow humans, and God according to their intrinsic value and unique role; moral sensitivity reinforces our accountability for the welfare of all we touch. Disregard for the prerogatives of relationships diminishes and eventually destroys the offenders.

The environmental crisis is not caused primarily by too many people, too little food, or too advanced technology. Environmental disaster looms because too many "managers of Earth" have abused the rights of others. Too few "lords of creation" have followed the example of Jesus who said, "Whoever wants to be the great man among you must be your servant, and whoever wants to be first among you must be the slave of all; just as the Son of man has not come to be served but to serve" (Matt. 20:26-28, Moffatt).

Humanity's Homeland: Earth's neighborhoods stretch around the globe, as seen in this Apollo 17 spacecraft view, courtesy of National Aeronautics and Space Administration.

COMMUNITY LIFE

Community is the essence of healthful life: interacting; interdependent; interdisciplinary. And man makes the intricate, controlling choices that promote or impair harmony within human and natural communities. The choice is religious, moral, and spiritual. "I have set before you life or death, blessing or curse: Oh, that you would choose life, that both you and your children might live," declared Jehovah in Deuteronomy 30:19, (TLB).

Since the turn of the century, science has corroborated many more of Scripture's basic views. When four billion people joined the world's foodlines and scientists informed us that the marketing of fluorocarbon aerosol products might destroy the ozone shield and turn sun rays to death rays, mankind became more aware of the answer to Cain's question: "Am I my brother's keeper?" (Genesis 4:9). And in the spotlight of ecology some people have verified Ecclesiastes 11:1, "Cast your bread on the surface of the waters, for you will find it after many days."

The world is a community, and our individual decisions cause wide-ranging effects. According to biblical cosmology, the consequences of our actions in the natural, visible community reverberate in a spiritual sphere as well. The rewards for observing these laws of the universe are both spiritual and material, Jesus declared. "Give, and you will have ample measure given you—they will pour into your lap measure pressed down, shaken together, and running over; for the measure you deal out to others will be dealt back to yourselves" (Luke 6:38, Moffatt).

Some advocates of "lifeboat ethics" declare that sharing the world's limited goods can only compound the crisis of shortages. "The fundamental error of the sharing ethics is that it leads to the tragedy of the commons," asserts biologist Garrett Hardin.[22] In this analogy, the size of the globe's resources, or "commons," remains fixed while the number of people using them increases continually. Some people must be denied access to the commons, Hardin says, or the result will be disaster for everyone.

But it is unneighborly exploitation, not cooperative sharing, that has shrunk world resources to the danger point. One part of the world cannot isolate itself from the rest which supplies its raw materials—or manufactured goods—or food—or cultural riches—or whatever value it needs and lacks in itself. Rather, all humanity can progress by recognizing its interdependence and acting on a Golden Rule ethic that promotes our neighbors' interests along with our own. "Whatever you would have men do to you, do just the same to them," Jesus directed (Matthew 7:12, Moffatt).

It is apparent that mankind hungers for more than bread; yearns for something more satisfying than affluence. Joseph Sittler, the ethics theologian, writes:

Is it possible that the Creator-Word, by whom all things were made, should be driven from his field by us? The central assertion of the Bible is that he has not been so driven, but rather drives, loves, and suffers his world toward restoration. It is of the heart of the Christian faith that this mighty, living, acting, restoring Word actually identified himself with his cloven and frustrated creation which groans in travail. "The Word became flesh and dwelt among us." To what end? That the whole cosmos in its brokenness—man broken from man, man in solitude and loneliness, broken from holy communion with his soul's fountain and social communion with his brother—might be restored to wholeness, joy and lost love.[23]

In the hands of people who care about others, modern technology can be a humanitarian tool. As philosopher Jerry Gill points out, technology is a "part of the created order, carrying with it the possibility of freedom from hunger, pain, ignorance, and political oppression."[24]

That is the possibility for good stewards of Earth. It's the law: "Whatever you sow, this shall you reap."

FIVE
SWITCHING
POWER

The greatest nation in the world—leader in scientific achievement, industrial production, and military strength—faces a momentous problem: a scarcity of the energy that helped power it to preeminence. More than any other nation, the United States depends on prodigious quantities of energy to create climate-controlled enclosures, power industrial operations, transport people and goods, and produce food. Once-cheap fossil fuels are moving toward the precious commodity category, forcing policy makers in energy, finance, and politics to make crucial decisions about power sources for the future. Their actions not only affect the availability of energy ten years from now but the health and living standard of energy users tomorrow. But the experts disagree on national energy policies.

DECADE OF DECISION

In 1974 a $4 million Ford Foundation Report purported to chart the nation's future energy path. Titled *A Time to Choose*, the twenty-one volume report drew on the experience of a Harvard dean, an executive in the nation's General Accounting Office, the chairman of Aluminum Corporation of America, the executive director of the Sierra Club, the president of an oil company, and other specialists. But instead of achieving a consensus, according to editor Lewis Lapham of *Harper's* magazine, the members vigorously promoted their separate causes.

> (For) two years they carried forward what the foundation was pleased to call "a dialogue" in which their fear, prejudice, and anger frequently obscured the questions at hand. . . . People associated the advancement of their own interests with the preservation of Western civilization and the American Way of Life. Experts came and went . . . their statistics provided a screen behind which the interested parties could haggle about the division of the spoils.[1]

The dispute over policies, priorities, and prices prompted a scholarly refutation called, *No Time to Confuse*. Sponsored by the Institute for Contemporary Studies, ten academic and business authorities elaborately repudiated the arguments of the first group of specialists. It should be no surprise, then, that the general public had difficulty perceiving an energy crisis.

No one can claim to possess all the facts in this complex and urgent debate, but it is evident that judicious decisions require a broad perspective of various fields, a respect for the health and prosperity of future generations as well as our own, and a realization that all energy-environmental actions affect the biosphere for good or ill. Informed citizens concerned for the whole of life are the best guarantee of wise choices in the energy options facing humanity.

NONRENEWABLE POWER—GOING

Oil, coal, and natural gas provided approximately 96 percent of the energy used throughout the world in 1975 (excluding food energy and the indeterminate quantity of waste products burned for cooking and heating in the developing countries). The remaining 4 percent of energy, according to the United Nations *Statistical Yearbook* of 1975, came equally from hydroelectric plants and nuclear reactors. Although huge supplies of fossil fuels still exist underground, swiftly changing conditions in society signal the necessity of switching from nonrenewable power sources in the near future. The new energy sources may generate the greatest technological, financial, social, and environmental changes since the Industrial Revolution.

Oil. Oil-producing nations are capable of filling all customer orders today, but they are willing to do so only at exorbitant and constantly rising prices. Saudi Arabia, who earned more income from the interest alone on its 1978 financial investments than from its total oil sales in 1972, jumped its 1979 price for crude oil from $14-plus to $18-plus per barrel, and in early 1980 to $26. After counting $40 billion in 1976 oil earnings, the Saudis launched a five-year $142 billion modernization extravaganza for its six million people.[2] Striving to spend $1 billion allotted annually to education, it finances such items as frozen lunches airfreighted from Europe daily for 200,000 pupils.

Iran, another oil-rich kingdom, scheduled a $600 billion expansion program over ten years. The grandiose plan collapsed with the national revolution in 1979, but billions of dollars in oil revenues continue to flood the coffers of these desert nations.

Oil surpluses appeared on world markets when quadrupled prices temporarily curtailed energy consumption in 1973. Many observers concluded that the "energy crisis" was a myth. But the quantity of available oil fluctuates with shifting international conditions, creating a constantly changing picture of energy supplies. Closed gas stations and long lines of cars creeping toward open pumps in the summer of 1979 again confirmed the uncertainty of supply.

The hard facts are:

1. Rising standards of living in every country require increasing amounts of energy.
2. Oil provides nearly 50 percent of the energy now used.
3. Mideast countries possess 30-40 percent of world oil reserves and control steadily rising prices through the international cartel.
4. Eventual exhaustion of oil reserves necessitates a massive switch to new power sources.
5. Current decisions on power will determine the sources a decade from now.

American oil companies continue to search for new reserves on this continent, but with little hope of replacing large foreign sources. Whereas America paid $7.6 billion for imported oil in 1973, it spent $41.5 billion in 1977! Even if five more Alaskan-size fields were developed here, they would not make the U.S. independent of foreign oil.

Urgency in switching power is the key conclusion of a fifteen-nation study group that included energy, academic, and governmental representatives. Project coordinator Carroll L. Wilson, of the Massachusetts Institute of Technology, summarized their findings: "The free world must drastically curtail the growth of energy use and move massively out of oil into other fuels with wartime urgency. Otherwise, we face foreseeable catastrophe."[3]

To meet the challenge, most energy experts recommend increased use of natural gas, coal, and nuclear energy. But these fuels have significant liabilities as well as limited quantities. A growing number of energy specialists endorse expanded use of solar energy now and major dependence upon it in the future.

One of the voices calling for new directions is Amory Lovins. In his book, *Soft Energy Paths*, Lovins describes the social, economic, and geopolitical advantages of renewable energy over fossil and nuclear fuels.[4] Rapid development of the "soft energy" alternatives would achieve enormous gains for almost everyone—and even the fossil fuel producers would be busily employed during the transition era.

Natural Gas. Residential and industrial customers for natural gas in America have sought more of this clean fuel than could be supplied. Shortages and surpluses have alternated in various regions of the country under changing government regulations. New imports from Mexico or Canada may temporarily make natural gas one of the nation's best energy options, but it is strictly a short-term (two or three decades) source.

Natural gas has been vented and burned in the atmosphere as a by-product of oil production in many countries. Its rising commercial value, however, is spawning a new industry: gas liquefaction. But ocean transport entails high financial and safety risks.

72

Algeria, Saudi Arabia, and Iran have launched multibillion-dollar projects for liquefaction facilities and for double-hulled tankers that can transport the liquid natural gas (LNG). Whereas Boston has been the lone American port receiving liquefied gas from abroad, plans call for new terminals on the West Coast, the Gulf of Mexico, and the East Coast to import over one trillion cubic feet a year by the early 1980s.[5]

Controversy is mounting over the LNG development. Most industry and government spokesmen regard the processing and transport equipment as sufficiently safe. What they cannot guarantee, of course, is absence of accidents or sabotage that would release the super-cooled liquid from confinement into the air where a spark could touch off an inferno. Rupture of an LNG tanker at sea could evaporate the $150 million vessel, while a bad leak in an inhabited area could do untold damage to property and people. In addition, large imports of LNG would exchange one foreign energy dependence for another.

Shale and Tar. Oil shale and tar sands contain immense quantities of petroleum, but costs for extracting and processing the fuel from shale and sand on a large scale are prohibitive by present methods.

The energy density of this fuel is only one-fifth that of bituminous coal, and mining residues require more space than the original material, causing serious environmental problems. Alberta, Canada—which produces a small amount of oil from tar sands—and the Green River shales lode in Colorado, Utah, and Wyoming hold the largest deposits in North America.

Coal. Can intensified coal mining take up the slack? It can if quantity is the main criterion. Reserve coal supplies are sufficient to last one hundred years at a 5 percent annual increase.[6] Such an increment would approach the President's original energy goal of mining one billion tons a year by 1985, but formidable obstacles block this route.

Anthracite, the hard, clean-burning coal, makes up only 2 percent of the U.S. coal reserve. Much of the bituminous or "soft" coal, used extensively for generating electricity, has a high sulfur content that requires costly pollution-control equipment. Subbituminous and lignite coal are relatively free of sulfur but have high ash-pollution and the lowest heating efficiency. By paying more for lower-grade fuel, installing and operating still-controversial pollution control devices, and also upgrading the nation's rail transportation, it is possible to substitute coal for oil to a significant degree. But there are additional high costs in human and environmental terms.

Current coal mining processes disrupt land—and people's lives. Deep mining is the most hazardous industrial occupation in the country, and strip mining makes wastelands. Perhaps the people of Appalachia have suffered the most in this regard. Though the miners earn good wages, their health is imperiled by black lung disease, family stability is undercut by frequent moves, and their communities are jeopardized by flash floods.

In 1972 a coal company-constructed dam collapsed and released millions of gallons of slag-filled water down Buffalo Creek in West Virginia. Not until weeks later was the one hundred and twenty-fifth victim found, and seven bodies never were recovered. Four thousand of the area's five thousand residents were left homeless.

In 1977 another flood devastated Appalachia in four states—Kentucky, West Virginia, Virginia, and Alabama—damaging 18,000 homes and 1,600 businesses and inflicting major property losses on 225,000 families.[7] Coal country is notoriously inhospitable to homemaking since prosperity depends on displacing the surroundings and moving on to new territory.

In land-short and oil-dry West Germany, 18,000 residents were uprooted in a 500-square-mile area to permit coal extraction from beneath their homes. An additional 10,000 Germans are scheduled for resettlement as the nation strives to replace Mideast oil with its own coal.

Some of America's best farmland is being destroyed by strip mining in Illinois. Farmers are offered two to three times the going price of acreage by coal companies. At stake is a food production resource for years to come versus a one-time transfusion of coal for gaping furnaces.

New federal laws require strip-mining operators to restore gutted land to its original productivity "within a reasonable length of time," but confusion surrounds their workability. Several states have their own laws but Illinois farmers declare that full recovery of mined land is impossible, and Illinois state officials indicated they would continue to issue mining permits until reclamation research is completed some years in the future.[8] Thus strip mining regulations decorate the law books but do little for the environment.

Energy developers and environmentalists in western states have clashed furiously over control of wilderness regions. Ready granting of mining leases on federal (public) land awaits extensive survey results and congressional decision. The disposition of over 500 million acres of territory will be determined by the Federal Forest Service and Bureau of Land Management.[9]

Much more than scenic landscapes are involved in the environmentalists' goals. Coal mining causes soil erosion and silting of streams, unpredictable settling of land, acidifying of soil and waterways when coal-pyrite is exposed to air and rainfall, shifting of water supplies, and loss of crop land after strip mining. As the great bulk of coal reserves are deep underground, expansion of deep mining could avoid some of the environmental damage caused by surface operations. Health conditions in underground mines must be greatly improved, however.

By the mid-1970s more than 100,000 miners were victims of "black lung" or coal dust disease, with four thousand deaths annually traced to its effects. The federal government pays over $1 billion a year in medical compensation to confirmed victims, with the amount rising steadily.[10]

The huge coal reserves in America will be a major boon if technological advances can achieve breakthroughs in the gasification, liquefaction, or solvent refining of coal. These complex procedures reduce sulfur content and provide fuel that is less polluting than petroleum. Pilot plant testing will determine the commercial feasibility of these methods by the early 1980s.

Unless practical systems are devised to burn coal with less contamination of the atmosphere, it is unlikely to supplant petroleum as the primary fossil fuel in tomorrow's economy. The General Accounting Office, a watchdog committee on congressional finances, estimated that the government's proposed doubling of coal use would require a $19 billion outlay for pollution equipment plus a $1 billion annual operating fee.[11] A shift toward coal without development of economical pollution devices would generate strong pressures to lower federal pollution standards, deferring the cure of environmental ills once again. Human health and functioning ecosystems are as vital to society as its economic expansion.

NET ENERGY

In this energy-short era, a vital guideline for choosing among energy alternatives is the matter of *net energy*. Simply put, it takes energy to produce energy. So, it takes energy to mine coal, to refine petroleum, to operate nuclear power plants. Net energy is the measure of the energy created minus the energy consumed. Net energy is a calculation which helps avoid superficial, injudicious decisions regarding energy development.

When fossil fuels were plentiful and easily accessible, the energy consumed in the process of the discovering, mining, transporting, and converting of power yielded a substantial net energy for diversified uses in society. But increasing difficulty in providing energy has reduced the net gain and in some cases eliminated it. These shifts call for a thorough reassessment of the energy source.

MORE FOR LESS

A long leap toward energy *sufficiency* can be taken by increased energy *efficiency*. Better methods and technology will require less energy for the same effects. Better insulation of dwellings and commercial buildings is one such improvement. Though material and labor costs are higher for such structures, the price is often less than the production of additional energy, and the eventual operation savings permit expanded purchases in other areas of the economy.[12]

Window space and artificial lighting are excessive in many commercial buildings, causing heat waste in cold periods and excessive cooling burdens in the summer. A study by the

American Institute of Architects estimated that the incorporation of efficiency and conservation measures in new buildings could save the equivalent of 12 million barrels of oil each day by the year 1990.[13] (This includes such conservation steps as moderately lowering thermostats in cold periods and raising them in warm stretches.) Heating and cooling of buildings consumes approximately 20 percent of the energy used in the U.S., making this a major area for improvement.

Industrial efficiency may produce still greater savings. Industry uses almost 40 percent of the total energy consumed in the U.S. During the energy squeeze of the 1973-74 oil embargo, some large industries reduced waste and used 20 percent less energy while maintaining their production output.[14]

Additional large savings are possible through new processes and equipment. Cogeneration of electricity and steam could utilize heat that is wasted in the separate processes; the new basic oxygen process for steel production uses one-fourth the energy of the open-hearth process; and a new system for producing aluminum uses 40 percent less energy than the conventional electrolysis method.

Heightened efficiency of automobile engines has markedly improved the energy situation. Government regulations have forced auto manufacturers to achieve higher miles-per-gallon standards, lengthening the intervals between gas pump stops. Auto makers reduced vehicle weight and engine size to meet government regulations, and future down-sizing will multiply motoring miles while reducing gas consumption and air pollution.

Another jump in engine efficiency looms in the redesigning of the power unit. Laboratory tests at the California Institute of Technology have proved that the spark-ignition system in recent model cars is about 20 percent more efficient than early '70 mechanisms.[15] Also, the fuel-injected, stratified-charge design pioneered in U.S. army vehicles in the late 1960s gives a significant boost in gas economy. The latter innovation also diminishes pollution by using a leaner mixture of fuel and igniting it at a higher ratio of air to fuel than in conventional engines, producing more power through a more complete combustion of less fuel.

LESS IS BETTER

Improved efficiency sometimes means reduced performance, such as less-powerful auto engines, lower-temperature "hot" water in homes, and fewer automatic high-energy conveniences. These savings stem from actual conservation of energy—a nominal down-grading of living standards that slashes energy consumption drastically on a nationwide scale. When socially conscious Americans recognize the depth of the energy-environment crisis, they will regard flagrant waste of energy as an extravagance we can ill afford. A study by the National Research Council, partially funded by the federal Department of Energy, concluded that living standards can continue to rise while energy consumption is reduced if gross waste is curtailed.[16]

NUCLEAR POWER—COMING OR GOING?

Until recently, nuclear energy was viewed as the miracle deliverer for an energy-hungry world. In 1976 Edward Teller, the "father of the atomic bomb," asserted that only nuclear reactors can meet the energy needs of the world's projected seven billion people in the year 2000.[17] Teller, the first chairman of the Atomic Energy Commission's advisory committee on nuclear safeguards, claimed that nuclear energy is essential for alleviating what he called "pollution by poverty" among the globe's masses. Washington Governor Dixy Lee Ray, a former AEC chairman, and radiation biologist, visualized a Niagara of power gushing from uranium and plutonium reactors. Predicting an increase from America's three-score-plus plants to between four hundred and one thousand, she predicted, "We will have all the electricity we need for a thousand years."[18]

But Teller, Ray, and other advocates of nuclear energy are losing many of their followers. From the president of the United States to increasing numbers of everyday citizens, resistance is growing to the proliferation of nuclear power plants. The dilemma was pinpointed by Alvin Weinberg, director of the AEC's Oak Ridge National Laboratory, when he stated, "We nuclear people have made a Faustian compact with society: we offer. . . an inexhaustible energy source. . .

78

tainted with potential side effects that, if uncontrolled, could spell disaster."[19]

While Weinberg continued to support the expansion of nuclear reactors with accompanying safety measures, many other scientists and economists asked whether the possible gains are worth the unavoidable risks, for nuclear energy is proving to be not only potentially more hazardous than expected but also more expensive.

About 12 percent of the nation's electricity in 1977 came from nuclear power plants. This was an energy equivalent of 425 million barrels of oil, but huge quantities of oil and coal were consumed to fuel the plants that produced uranium fuel. In fact, environmental writers Edward Flattau and Jeff Stansbury point out that in 1972 over 25 billion kilowatt-hours of electricity were used to produce the enriched uranium that fueled nuclear plants which had a total output of about 50 billion kilowatt-hours. Supported by calculations of E. J. Hoggman, a University of Wyoming nuclear specialist, the writers estimated the net gain from atomic energy that year was about 10 percent after all energy expenditures for procuring the uranium fuel were deducted.[20]

Howard Odum, an ecological engineer at Florida State University, has questioned whether any net energy has yet been obtained from the nuclear industry if all costs are put into the cost-benefit calculations.[21] Just one major reverse, such as the multimillion-dollar clean-up and reconstruction costs after the nuclear plant accident at Three Mile Island in 1979, sends net energy figures into a tailspin.

Between 1973 and 1976, the cost of producing nuclear energy increased dramatically. Tripling of the price of uranium ore, inflated costs for long-term construction and financing of nuclear plants, time delays for environmental impact studies, and steadily rising prices for the oil and coal used in processing raw uranium have greatly reduced or wiped out the economic advantage of nuclear power. Instead of producing electricity 80 percent of the time as projected, many plants have been shut down up to 40 percent of the time for repairs or inspection.[22]

Ironically, the energy shortage of 1973 crippled the nuclear boom instead of accelerating it. When OPEC tripled oil prices and Arab producers brandished an embargo threat over the world, the growth of electrical use in the U.S. fell from 7 percent a year to nearly zero—and utility companies began cancelling orders for reactor plants. In five years new construction orders fell from thirty-six a year to four, and the mushrooming $10 billion a year industry took on the aspect of a shrinking violet.[23]

The four major manufacturers of nuclear plants—Westinghouse, General Electric, Combustion Engineering, and Babcock and Wilcox—continued in business with some one hundred and fifty plants under construction or on the drawing boards. But plans for sixteen reactors have been dropped and sixty-six more have been deferred in the past three years, decreasing anticipated construction far below the thirty-plant annual capacity. Only two plants were ordered in 1978.

During the slump in reactor construction, troubling disclosures about nuclear operations began seeping into public consciousness. A 1969 study showed that uranium miners were suffering lung cancer rates—from radon—about four times as great as the rest of the population.[24]

In March 1975 the world's largest nuclear plant, at Brown's Ferry, Alabama, overrode several safety systems that protect against accident when a fire caused considerable damage to the electrical power systems. If a backup apparatus had not functioned, the uncooled uranium core could have melted its container, split open the plant, and released radioactivity into the atmosphere that would have killed nearby people and produced cancer and other serious health problems in countless others.

Pennsylvanians experienced nuclear jitters in March 1979 when equipment failure and human errors at the Three Mile Island reactor near Harrisburg released radioactive steam and damaged controls of the core cooling system. The accident exposed ominous gaps in the industry's emergency capabilities, and the "unprecedented contamination" of the facility may keep it unproductive for years.

Prior to Three Mile Island, the more serious risks of radiation damage had shown up at facilities handling plutonium for fuel-reprocessing or bomb-making farther back along the multi-stage nuclear-production line.

Uranium is extracted from sandstone ore at mills, leaving radioactive sandstone residue, called tailings, the first disposal problem in the nuclear-power trail. Since the resulting uranium oxide contains only a small amount of the uranium-235 that is fissionable in the U.S.-style Light Water Reactors, the U-235 must be enriched. To accomplish this, the uranium oxide ("yellow cake") is first changed to gaseous hexafluoride at conversion plants, then sent to gaseous diffusion plants in Oak Ridge, Tennessee, or Paducah, Kentucky, or Portsmouth, Ohio. These plants were built in the 1940s and 1950s to supply high-level enriched uranium for nuclear bomb production as well as low-level uranium for power plants. (Canada's Heavy Water Reactor burns uranium-238 without fuel enrichment.)

The costly and complicated enrichment plants have been the main check on global expansion of nuclear power, but Canada and several European nations have produced alternate systems and have signed contracts to construct nuclear facilities in other countries, breaking the United States' near-monopoly. Meanwhile, technicians in a number of countries, including India, have learned to refine plutonium from used reactor fuel and subsequently built their own bombs, increasing the international nuclear stockpile.

For reactor use, fabrication plants package the enriched fuel into pellets encased in rods. A bundle of rods constitutes the fuel core that is placed in the reactor vessel of a power plant. As the fuel is activated, producing heat, steam, and electricity, one-third of the core is withdrawn each year and replaced with fresh fuel. The spent fuel is then stored or sent to a reprocessing plant for further refining.

Reprocessing is a strategic and increasingly controversial stage of nuclear power production. This step conserves uranium ore by recovering uranium-233 and recycling it by gaseous conversion, enrichment, and fabrication for reactor cores. But reprocessing also yields highly radioactive waste material which must be stored, plus pure plutonium which can either be refabricated for reactors or forwarded to nuclear bomb plants to comprise the "trigger" on hydrogen weapons. Security against theft of plutonium by terrorist groups is also a significant factor in the reprocessing decision.

West Valley, New York, the Buffalo-adjoining site of the

only "successful" reprocessing plant in the U.S., dramatized the quandary of nuclear power advocates. There, Nuclear Fuel Services, a private company operating under AEC auspices, reprocessed nuclear fuel between 1968 and 1972. Then the government ordered the plant remodeled for greater protection against earthquakes, tornadoes, and harmful radioactivity. (Two other reprocessing plants in Morris, Illinois, and Barnwell, South Carolina, never began operations because of technical and political obstacles.)

What began as a $40 million remodeling project at West Valley turned into an estimated $600 million-plus reconstruction, and Nuclear Fuel Services cancelled further operation plans. It announced it would return the facility to the government when its lease expires. But the worst problem remained: the safe disposal of 550,000 gallons of dangerously radioactive waste stored in an underground tank.

Congress allocated a million dollars to investigate ways to dispose of the radioactive sludge and to decommission the defective plant. Argonne National Laboratory in Chicago got the assignment. Dr. Philip Gustafson, an Argonne scientist, said the project might cost from $30 million to $600 million, adding: "If this is done properly, it removes the biggest obstacle to the public being reasonably comfortable with nuclear energy."[25]

It's a big "if." After two decades of nuclear processing, the government has not found a fool-proof method of disposing of wastes that will be radioactive up to 100,000 years. Man-made plutonium, the key ingredient of nuclear bombs and also the main fuel for future-stage breeder reactors, is so toxic that one-millionth of a gram ingested by a human will cause cancer.

The uncompleted reprocessing facility at Morris, Illinois, shelters four hundred times as much highly radioactive material as the West Valley site. Here the waste is in the form of spent fuel rods collected from reactor plants across the country. Readily transferable in rod form, the 310 tons of fuel rest in a concrete-walled pool awaiting government decision either for recycling at a reprocessing plant or for storing permanently—if safe disposal plans can ever be made.[26] Meanwhile, lower-lever radioactive waste in barrels is filling up some dump sites.

Plutonium danger caused wide consternation at Broomfield, Colorado, a suburb of Denver. The nearby Rocky Flats factory has produced triggers for nuclear weapons since 1951, with radiation accidently released several times. The most alarming mishap deposited a layer of plutonium in Broomfield's drinking water reservoir. An EPA report in 1975 confirmed that seepage from buried radioactive wastes had been carried by a stream near the plant to the Great Western Reservoir. Residents were reassured that the plutonium's weight kept it on the bottom of the supply basin, but many of Rocky Flat's neighbors say the plant has worn out its welcome despite its annual infusion of $47 million into the local economy.[27]

Emigrant scientists from Russia recently confirmed that a nuclear accident leveled a wide area in 1957 and probably killed thousands of their countrymen. Soviet authorities suppressed the news and even Western officials avoided the sensitive subject. But biochemist Zhores Medvedev and physicist Lev Tumerman, exiles to England and Israel, respectively, revealed in 1976 that a nuclear disaster struck the area around Kyshtym in the southern Ural Mountains, long a military-industrial complex.[28]

Writing in *The Jerusalem Post*, Tumerman blamed the Russian debacle on an explosion of radioactive wastes, agreeing with Medvedev on the cause. Tumerman drove through the devastated area in 1960, and later recalled, "Only chimneys remained of towns that once were there. As far as the eye could see, there were no villages, no towns, no people, no cattle herds." Thus nuclear wastes under some conditions constitute not only a radiation danger but an explosive hazard of significant proportions.

Safe containment of radioactive waste is a growing dilemma for the nuclear power industry. Storing low-level uranium waste in steel containers and submersing high-level waste in water are only short-term and inadequate remedies; the long-term disposition of these materials will affect the health and even existence of future generations.

Restraining the nation's rush toward plutonium dependence, President Carter announced in April 1977 that the government would postpone development of reprocessing plants that refine plutonium. In addition, said the President, the U.S. would delay research on breeder reactors, the plutonium- and uranium-fed power plants that will theoretically produce more plutonium than they consume—for either refueling power plants or making nuclear weapons. Instead, the U.S. will supply its licensed reactors here and abroad with low-level enriched uranium.

The President explained his decision as a brake on the production of bomb-grade plutonium which is easily convertible to nuclear explosives. The decision also delays the nation's technological leap into the next generation of reactors —the "fast breeder" model that uses plutonium along with uranium-238 and thorium-232 to produce extra plutonium for other reactors. Operating in tandem with thermal reactors, the breeder would greatly extend supplies of uranium.

Breeder reactors, however, have not yet proved themselves. Two in Western Europe and one in Russia operate with some success, but the experimental Clinch River breeder plant near Oak Ridge, Tennessee, faces an uncertain future. After $500 million was spent for design and parts, no construction had begun, development schedules were delayed eighteen months, estimated costs had tripled to $3 billion, and the President and Congress were contradicting each other over breeder feasibility and strategy.[29]

U.S. decisions on nuclear power development have worldwide ramifications. Although Russia built and services reactors in Communist-bloc countries, and German, French, and Canadian manufacturers are competing for reactor sales to other countries, the United States has built and licensed nearly three-fourths of the two hundred-plus reactors around the world. These power plants in nearly twenty countries are dependent upon U.S.-supplied enriched uranium, and even the breeder-operating countries will be influenced by the progress of still-developing nuclear technology in the U.S.

Many scientists and government leaders had assumed that breeder reactors would succeed present thermal models. Everyone knew uranium fuel is limited, and something would have to take its place. But the tremendous expense of developing breeder technology, the unmeasured hazards of nuclear processes, and the figurative plutonium cloud shadowing world horizons have turned nuclear escalation into a gigantic question mark.

Even if breeder reactors are not perfected and scattered around the world, the slow production of plutonium as a by-product in conventional reactors will still make bomb manufacture possible in a score of countries within a few years. The superpower nations may keep their own finger off nuclear triggers but not be able to control the actions of such countries as Israel, China, South Africa, and Iran. They too have nuclear bombs or will soon have bomb-making capacity. Israel reportedly was readying nuclear weapons in the 1973 Mideast war just before the tide turned in her favor. And Russia appeared to be joining the Arab defense until an American warning cooled the war fever. The wider the use of nuclear energy, the greater the danger of nuclear war—or of nuclear blackmail by terrorists.

Already significant quantities of uranium have been "lost." In the mid-1960s more than 400 pounds of U-235 could not be accounted for by the U.S. government, though even small amounts are monitored by law. Investigation failed to locate the bomb material.[30]

In 1968 more than 200 tons of milled uranium disappeared from Europe. The nuclear division of the European Economic Community, EURATOM, lost track of the huge shipment in a flurry of paper transactions, cargo transfers, and international intrigue that baffled, bribed, or intimidated nuclear guardians. The $3 million acquisition supplied an unknown nation—identified as Israel by some speculators—with unrefined uranium that could be processed by knowledgeable technicians into fissionable explosives. And nuclear war moved a step closer to ignition.[31]

For energy and for warfare, nuclear technology is a schizophrenic servant, a treacherous ally. When out of control, nuclear energy dooms humans within its reach.

The unparalleled jeopardy of a nuclear society was wryly summed up by Nobel Laureate physicist Hannes Alfven. "Fission energy is safe only if a number of critical devices work as they should, if a number of people in key positions follow all their instructions, if there is no sabotage, no high-jacking of the transports, if no reactor fuel processing plant or reprocessing plant or repository anywhere in the world is situated in a region of riots or guerrilla activity, and no revolution or war—even a 'conventional' one—takes place in these regions. . .no 'acts of God' can be permitted."[32]

Publicity about nuclear power in Britain temporarily halted that nation's expansion of breeder reactors. While the chairman of the Atomic Energy Authority, John Hill, expressed confidence that all the nuclear problems could be solved, a distinguished study commission urged caution and judicious public evaluation.

Led by physicist Brian Flowers, former director of the government's atomic research headquarters, the study concluded: "There should be no commitment to a large nuclear program including fast reactors until the issues have been fully appreciated and weighed in the light of public understanding. . . . We should not rely for energy supply on a process that produced such a hazardous substance as plutonium unless there is no reasonable alternative."[33]

Is there a "reasonable alternative"? Renewable, inexhaustible energy sources such as solar, wind, and biogas are not only a reasonable alternative but are within the reach of today's scientists and fiscal planners. Only the national decision and commitment are lacking.

Energy specialists have typically projected solar and other renewable power systems for the distant future. National budgets have allocated token funds for solar research while lavishly supporting nuclear development. Nobel prize winner George Porter trenchantly observed: "If sunbeams were weapons of war, we would have had solar energy decades ago."

But perspectives and policies are beginning to shift. As nuclear energy reveals its insidious liabilities, a growing number of scientists, politicians, and environmentalists are encouraging all-out development of safe, renewable energy sources.

RENEWABLE POWER—COMING, BUT WHEN?

Nature surges with power that could banish the energy crisis if human ingenuity can capture and channel it. The radiating heat of the sun, propelling force of the wind, expanding thrust of subterreanean steam, and rushing momentum of waterfalls and tides offer unlimited supplies of clean energy through enterprising technology. In addition, the spread of biogasification plants and development of tree farms for fuel promise to revolutionize energy production.

The Sun. Solar energy is the brightest option in the power switches. Celebrating the first Sun Day in May 1978, Colorado Senator Gary Hart told a crowd at the Lincoln Memorial in Washington, "Solar energy is environmentally benign, it is labor intensive, it is decentralized, and it can be individually owned and operated. It has all the elements that an awful lot of people can get behind."

Governments in over thirty states agree with Senator Hart to the extent of giving income-tax deductions to homeowners and businessmen who install solar equipment. The tax write-off went nationwide in late '78 as Congress approved a substantial credit for addition of home solar units and another for home insulation improvements. This monetary inducement will stimulate growth of a new industry as well as saving immense amounts of fossil fuel over a period of time.

After the oil embargo in 1973, a variety of solar energy businesses entered the marketplace and by 1977 had boosted annual sales to $125 million. Expectations jumped in 1978 with the government announcement that solar systems had become economically competitive with electric space heating systems and that solar water heaters could feasibly replace conventional units.

Industrial development of solar equipment has been erratic because of high costs and widespread indifference to sun power. The economic blueprint took on a golden hue, however, with new policies of tax credits, a near-tripling of government-funded research for 1979, and the prospective removal of price ceilings on gas and oil which have subsidized extravagant use of fossil fuels and consequently handicapped alternate energy development.

When solar systems eventually replace fossil fuel furnaces in homes and commercial buildings, an enormous task will still remain. The challenge is to produce photovoltaic cells economically that convert sunlight directly to electrical current.

Photovoltaic cells now power small orbiting spacecraft, drawing continuously on the sun's ray. Composed of silicon crystals (or other light-sensitive materials), solar cells currently are expensive to manufacture and have a somewhat uncertain life-expectancy. A storage system or battery for the generated power is also insufficiently developed.

Among many others, engineer S. David Freeman is optimistic about solar prospects. As chairman of the board of directors of the Tennessee Valley Authority, and former energy consultant to the federal government, Freeman voiced his enthusiastic outlook during a Sun Day rally in Memphis. Speaking in the vicinity of the largest utility and coal-burning plant in the country, Freeman declared that sufficient use of solar energy could avoid the building of a nuclear power plant. Barry Commoner reports that mass production of photovoltaic cells for Defense Department use could shrink costs from $15 per watt to 50¢, making the cell commercially feasible.[34]

Short of a big switch to solar, nuclear energy expansion will likely continue. Holland's agreement to sell uranium to Brazil for eight new reactors planned for 1981 operation is one signpost. U.S. commitment in 1978 to sell $1 billion worth of enriched uranium to Japan is another. Though atomic energy is no longer cheap and its use is increasingly demonstrated to be hazardous, its spread is guaranteed in the absence of a suitable substitute—unless public opposition should shut off the supply at its multiple sources.

Denis Hayes, former Worldwatch Institute chairman and originator of Sun Day, estimates that nearly all energy needs could be met by sun power within fifty years if priority is given to solar technology. He includes in his calculations such indirect solar forms as hydroelectric and wind power, and biomass fuel supplied by vegetation and wood. Biomass could provide one-third of world energy needs, Hayes suggests.[35]

Wind, Water, and Organic Matter. Federally funded research for alternate energy sources is increasing steadily—from $33 million in 1975 to a proposed $300 million in 1979. Scientifically designed windmills that generate electricity are in the forefront of experimentation.

Small-bladed windmills are tested by the Department of Energy at Rocky Flats, Colorado. The most efficient designs will be promoted for farms or homes to pump water or generate electricity. Louis Divone, head of DOE's wind systems branch, says: "I'm quite optimistic that the wind systems will be used as much in the future as they have in the past."[36] There were six million windmills in rural use in the 1800s.

Another test program operates large and innovative-style blades for high power output. Early results are encouraging, indicating that a large windmill could supply electricity for several hundred homes or that a chain of windmills could feed power to utility companies ordinarily operating on oil or gas. The system is already economically competitive, according to Ugo Coty, technical manager of wind energy studies at the Lockheed-California Company in Burbank. "Construction of two hundred and sixty large-scale windmills at sites having an average wind velocity of 15.7 miles an hour could bring the cost of wind energy down to the present price of oil," he says.[37]

The Western Plains area has winds between fifteen and twenty miles an hour, with much of the land within the public domain. A string of windmills on the plains and hills might be a novel sight, but they would be positively beautiful in contrast to the slag heaps of mining operations—and 100 percent less polluting. For economy, simplicity of technology, and environmental compatibility, windmills may be the most promising early source of supplemental electrical power.

There are several other alternatives to oil, coal, and uranium fuel. Methane gas is readily obtainable from animal manure and vegetation. A simple process mixes water with animal (and/or vegetable) wastes, maintains temperature conducive to the decomposition activity of anerobic bacteria, and pipes off the resultant methane to gas lines or an engine driving an electric generator. The system is primitively employed in energy-short, animal-rich countries such as India and China, but the tremendous potential is only beginning to be appreciated in the U.S.

Large cattle or hog farms have the capacity for generating

89

much of their power needs from private biogasification plants (illustrated in the next chapter). Not only can methane substitute for a farm's purchased oil, gas, or electricity, but the residue forms a rich fertilizer-humus for conditioning croplands and saves expenditures for petroleum-derived fertilizers.

The Calorific Recovery Anaerobic Process company of Guymon, Oklahoma, sees the commercial potential of biogas. It plans a $3 million plant to process feedlot waste from 100,000 cattle into methane. The expected 1.5 million cubic feet of gas could heat 3,500 houses.

Bioconversion of selected crops and fast-growing trees is also under development both here and abroad. Sugar cane and corn can yield distilled alcohol, and crop-fuel plantations may become industrial farms of the future. Brazil produced 200 million gallons of alcohol from sugar and manioc in 1977 and mixed it with gasoline for use in modified combustion engines. The methanol mixture burns cleaner than gasoline and gives better performance. The conversion of crops to fuel should be weighed against the needs of people for food, but the potential for crop-fuel from relatively inedible plant parts deserves careful investigation.

Another clean and relatively inexpensive energy source is heat from the earth's interior. Where such heat contacts subterranean water near the earth's surface, pressurized steam results. Geysers form where this steam bursts through the crust to the atmosphere. Capture of this geothermal energy for the powering of steam turbines produces electricity in many parts of the world. Notable geothermal fields exist in New Zealand, Italy, Iceland, and the United States.

The Geysers region near San Francisco supplies more than half of the electrical generating power of that city. Explorations and new production are rapidly expanding the Geysers' output, and major geothermal discoveries have been recorded in California's Imperial Valley and in areas of Utah.

Central American countries are planning greater production of geothermal power from wells in their regions. The presence of numerous volcanoes in this area heighten expectations for successful drilling. Overall, geothermal energy can be only a small part of needed energy expansion.

France, Russia, and China operate experimental tidal plants in offshore currents, but the U.S. has done little in developing tidal energy projects.

POWER FOR THE PEOPLE

Energy is an increasingly vital element in our national well-being. High employment depends on adequate energy supplies. Food production and distribution require enormous amounts of energy. Raising the living standards of poorly housed and undereducated people demands greater energy investment than in the past. Yet human health, as never before, is affected by the pollution factor of energy. We cannot expect energy to be cheap again, but it must be abundant, clean, safe, and decentralized to meet the needs of all the people.

Nonrenewable fossil fuels fail on two of the four criteria: abundance and cleanliness. And they are controlled by monopoly agencies, which are often deficient in responsiveness to the public.

Nuclear energy qualifies in only one respect: abundance. But that abundance magnifies the potential perils from nuclear radiation, susceptibility to terrorist threats, and Orwellian control of energy by remote and isolated bureaucrats. With nuclear power proliferation, the "man in the street" will see financial and political power moving beyond his control into a commercial-governmental oligarchy.

Government of the people, by the people, and for the people in a high-energy era needs the accessibility available in solar, wind, and biogas power. These renewable power sources also meet the criteria of cleanliness and safe use. Further, they create more jobs for workers than does the sophisticated technology of nuclear plants.

Nuclear energy was developed by scientists in an all-out effort by the U.S. to win World War II. Their success helped the democratic nations win the war, but they unleashed a force that may accidentally—or designedly—destroy the world. Today a crisis greater than international warfare may be near—the radioactive poisoning of the environment and humanity. The genius and wealth of America are fully as capable of developing safe energy sources as constructing doomsday weapons, and the prize is not only freedom but life.

The urgency is great to implement a comprehensive energy program. Current decisions about energy choices will affect conditions now and far into the future. Consumer-voters in America have the opportunity to help decide whether nature's power will be exploited for today's energy-elite or invested for tomorrow's energy managers. The next chapter documents some promising power developments that are well underway.

SIX
NEW HORIZONS

Some people insist on seeing a new phenomenon before they will believe in it. Other people believe in the possibility of unprecedented events and help make them visible to doubters. Here and there, perceptive individuals and organizations are uniting scientific inventiveness with ecological insight to form new vistas of progress. Their success demonstrates that economic expansion and environmental enhancement rise or fall together.

THE MUSKEGON MODEL

The Muskegon County Board of Commissioners in Michigan was one of the first public bodies to adopt a multipurpose plan for wastewater management rather than expanding its flush-sewage-down-the-river treatment. Revamping their initial goal of reducing phosphate discharge into local waterways, the county's twelve municipalities established a system that simultaneously advances agricultural production, encourages industrial-economic growth, restores recreational waterways, and reconditions wastewater to drinking-water quality. These gains were accomplished at a cost approximating that of building a secondary stage sewage plant, but the operating costs are half as high.

The land treatment plan for Muskegon County drew on

Purified effluent leaving the Muskegon site being examined by environmental group. This cool, crystal-clear manmade river supports a trout population.

successful experiments at Pennsylvania State University.[1] The Pennsylvania project had sprayed treated sewage on crop land and woods, increasing underground water supplies, tripling hay and corn crops, and accelerating the growth of trees.

Following resource management principles, the Muskegon plan treated sewage pollutants as resources out of place, recognized that pollutants are not usually destroyed but only confined or relocated, and improved the soil and air quality while renovating the water. Steps in the process are illustrated in the accompanying photos.

The project price tag—shared by federal, local, and state funds—was $44 million, including the purchase of 10,000 acres of land east of Muskegon and funding for a five-year research program, including a socio-economic study. Three eight-acre lagoons were constructed to receive the area's domestic and industrial sewage. Large enough to hold three days' intake, they allow time for bacterial colonies to decompose the domestic sewage and industrial waste, even when heavy rains heighten the inflow. This capacity is lacking in many conventional sewage plants, forcing operators to discharge improperly treated sewage into watercourses periodically.

Public officials viewing irrigated areas at the Muskegon County wastewater management site.

From the mechanically aerated lagoons, effluent flows to one of two eight-hundred-and-fifty acre storage basins where the nutrient-laden water accumulates throughout non-irrigating periods. In the growing season, giant irrigation rigs receive the wastewater and gently spray more than 5,000 acres of barren land—and excellent corn crops result.

Underground plastic pipes and a series of wells throughout the irrigated area carry away excess water and prevent salt buildup in the soil. The wells also provide stations for monitoring the soil-filtered water for purity level. Certain disease-causing viruses travel in water, but soil filtration is one of the best protections against viruses. Oppositely charged viruses and soil particles attract and hold one another until soil bacteria decompose the viruses into harmless protein. This health bonus is unique to the land treatment system of sewage disposal.[2]

The Muskegon project was acclaimed one of the top ten engineering achievements of 1972 by the National Society of Professional Engineers. It also gained attention in Washington where Congressman Guy Vander Jagt noted the significance of the enterprise and secured critical support for the project from the White House, the President's Council on Environmental Quality, and the governor of Michigan. Later he ventured: "I will predict that the Muskegon County facility will be the focal point for the nation's battle to solve water pollution problems."

Muskegon County's environmental transformation influenced congressional committees in formulating the Federal Water Pollution Control Act Amendments in 1972.[3] The legislation promoted community development of land treatment facilities, but wavering enforcement of the new regulations and deep-rooted opposition to changing the old system perpetuated waterway pollution in the following years.

Five years later, with water pollution costs threatening to submerge the national budget, Congress responded with stronger legislation in the Clean Water Act of 1977.[4] This law directs EPA to promote consideration of alternative methods of sewage disposal, such as land treatment, and prescribes 85 percent in federal funding support for alternative recycling systems rather than the 75 percent granted conventional sewage plant systems.

The federal funds offer Americans their greatest opportunity yet to clean up waterways, restore land fertility, reduce monumental fertilizer costs, and build associated energy-conserving industries. But the choice between the old, established method of sewage disposal and the new, multi-productive technology remains the option of local governments.

Water issuing from the "living filter" of suitable soil would not need the costly additional process of carbon filtering being advocated by EPA for the purification of drinking water in larger municipalities. The greater the population and industrial concentration, the more complex and expensive becomes the providing of potable water. Unless revolutionary methods are implemented, safe drinking water may become unaffordable for millions of people in this country.

Many communities have nonproductive land nearby that could be transformed into a water-revitalizing, soil-renewing, energy-conserving showplace of environmental management. The public leaders of Muskegon County have proved it can be done.

The Task Force Report on Water Pollution by consumer advocate Ralph Nader agrees: "There is ... one element of the

Muskegon County experience which unquestionably holds true for the country at large. We will only begin to outgrow our fixation on the halfway treatment technology handed down to us by the pollution control profession when we stop being satisfied with the moderately polluted rivers and lakes which that technology offers and begin to demand water that is truly clean."[5]

Land treatment, the "living filter" system, is the only feasible method for achieving the nation's zero-pollution discharge goal for waterways by 1985. However, legislators, water works lobbyists, and sanitary engineers continued to advocate advanced-stage sewage plants after passage of the Federal Water Pollution Control Amendments Act in 1972. Unsatisfactory results in city after city have subsequently proved the failure of conventional sewage treatment.[6] Even if chemical treatment were effective, the astronomical costs would cripple such projects before waterways became clean again. We must cooperate with nature's efficient system, or be satisfied with contaminated, potentially harmful water.

LUBBOCK, NORTHGLENN, AND NEW VISTAS

The long-established land treatment program of Lubbock, Texas, is now expanding to meet the needs of its growing community. For forty years, nearly 3,000 acres outside the city have produced cotton, soybeans, wheat, and alfalfa with the aid of treated sewage from the city. As the groundwater table sank rapidly and domestic sewage increased from the growing community, Lubbock recognized the opportunity to use its water efficiently while enlarging its sewage capacity. The result is an agricultural-academic-civic project that transforms a problem into an asset.[7]

More than 4,000 additional acres of thirsty farmland will receive municipal wastewater in the expanded operation. Fed by sewage nutrients and water, diversified crops will flourish in the dry highland area where cotton sometimes struggles to survive. The LCC Water Resource Institute, in conjunction with the LCC Investment Corporation, owner of the sewage-irrigated land, will conduct ongoing research at the new project in collaboration with the High Plains Water Resources Center and various departments of nearby Texas Tech University. The people of Lubbock have learned that sewage is no problem when properly managed.

Varied land treatment programs can be developed to meet local conditions. Northglenn, a suburb of Denver, devised a plan with surrounding farmers in 1977 that creates a fruitful new partnership. Instead of continuing to depend on a neighboring city for water and sewer service, Northglenn contracted with an association of farmers to first pipe their allotment of irrigation water from the mountains to city consumers. In turn, Northglenn would construct facilities to collect its sewage and storm water, treat it, store it in a lagoon, and return the nutrient-enriched water to the farmers for crop irrigation.[8] A shining example of urban-rural cooperation, the plan was praised by state governor Richard Kamm as "the most innovative, creative thinking I have seen in my ten years of politics." Also, President Jimmy Carter said, "This might be a vista of what we will see on a broad base in the future."

VIEW FROM MOUNT TRASHMORE

Facing the need for additional trash disposal space, DuPage County in Northern Illinois created a recreational marvel instead of an unsightly dump. Starting with an abandoned gravel pit, county officials created a lake and adjoining ski hill for community recreation activities.

Under the direction of county Forest Preserve leaders, gravel extracted to deepen the pit area was sold to commercial users; clay dug from the pit became "walls" that encased rising layers of garbage deposited by scavenger trucks. Scavenger company fees and gravel sales helped pay the costs of building "Mt. Trashmore," one of the country's first scientifically planned solid-waste mountains. Today the Blackwell Forest Preserve offers 80 acres of lakes for swimming and sailing, picnicking and camping areas, horseriding trails, and winter sliding on the 150-foot-high hill. The county is building additional trash-sculpted mountains for recreational use, and these second generation hills are designed to facilitate the collection and sale of methane gas from the decaying core of garbage.

Mt. Trashmore: Beauty and year-round recreational facilities highlight the manmade mountain and lakes fashioned from garbage at a gravel pit in the Blackwell Forest Preserve, DuPage County, Illinois.

The perspective from Mt. Trashmore is exhilarating. Trained eyes can envision similar recreational, scenic, energy, and water control benefits duplicated across the country. Every community has garbage, and garbage is a plus-factor for communities that have eyes to see wastes as resources. If "mountain building" is not suitable for a given area, recycling of trash would benefit communities rather than adding an unsightly dump. Recycled paper, wood, glass and metal would save vast amounts of energy and raw materials if nationwide collection and processing policies were adopted. Garbage and other organic rubbish could produce methane gas for heating homes and cooking food. Under proper conditions, decaying organic matter generates the type of gas that suppliers now transport great distances to meet consumer needs.

ANAEROBIC POWER TO BURN

Scattered across the country on countless livestock farms and feed lots are fuel reserves currently being ignored. They consist of large quantities of manure from animal herds and flocks. Under controlled conditions in a biogas plant, or anaerobic digester, animal wastes can produce methane gas to power most of the farm equipment, heat farm buildings, and provide nitrogen-rich fertilizer and soil conditioner for croplands.

Mason Dixon dairy farm, near Gettysburg, Pennsylvania, has such a plant. Designed specifically for this farm, the system includes a manure digester for six hundred dairy cows, a plastic bubble for storage of biogas, a storage facility for residual fertilizer, pumps for moving the manure, and a biogas-powered diesel engine which drives an electricity generator. The generated electricity meets about 60 percent of the energy needs of the dairy operation. The digester residue is a superior, odorless fertilizer ready for spraying on cultivated fields.

As in garbage-methane operations, rapidly multiplying bacteria are the prime movers of biogas power. They flourish in warm surroundings, changing the organic materials into methane, carbon dioxide, and water, and produce new colo-

1. MAIN HOUSE
2. MILK PARLOR
3. CALF BARNS
4. COW BARNS — 740' LONG
5. DIGESTER
6. POWER GENERATION BLDG.
7. BOTTLING FACILITY

MASON DIXON FARMS
EST. 1750

Up to Date: This historic farm in southeastern Pennsylvania produces its electric power for farm operations with a custom-built biogas plant powered by dairy cows' manure-to-methane energy.

nies when their food-fuel supply is sustained. In the process, the manure odor is removed.

Anaerobic bacteria do their work on the Mason Dixon farm in a large concrete box in the ground. Bacteria digest the manure for ten to twenty days to produce biogas (60 percent methane). The fuel can power farm vehicles by addition of a pressurized tank and a special carburetor valve for mixing diesel fuel with the gas.

Just as important as power to the farmer is fertilizer for his fields. Here too anaerobic bacteria excel in serving the farmer. Fresh manure loses some of its protein nitrogen to the air, sometimes causes pollution-runoff problems, and requires liming in fertilized fields to counteract acidity. But bacteria-digested manure avoids these liabilities. And the liquid form is easily piped to a storage pond where it awaits a convenient schedule for spraying on the fields in regular, efficient doses.

Anaerobic digesters can be built for different sizes and kinds of stock farms. Dairy and beef cattle, horses, hogs, chickens, and turkeys all supply manure for methane-gas and field-fertilizer when the animals are confined in feedlots, barns, or pens. Since farms use about one-fourth of the energy consumed

101

Design for the Hanson Park School in Chicago combines passive solar gain through south-facing windows, with an active solar system using solar collectors on the roof.

in the U.S., a significant shift to anaerobic digesters on stock farms would conserve huge amounts of both energy and chemical fertilizers.

Congress has encouraged biogas development by adding a 10 percent investment credit for such plants in the Energy Tax Act of 1978. This increases the investment credit for businesses to 20 percent for such facilities, making energy independence more attractive to livestock farmers.

TEXTBOOKS AND ENERGY

Environmental education in Chicago advanced significantly with the establishment of the Career Development Center for Environment and Energy Management Studies by the city's Board of Education.[9] The Board also contracted for the construction of Chicago's first sun-heated public school.

The Career Development Center links junior and senior high school students with Chicago businesses and agencies that produce, deliver, and regulate energy and manage the environment. The program allows students to "develop an understanding of career opportunities and the contributions of citizens to the improvement of the quality of life in Chicago," reports Donn Wadley, Director of Program Development for Alternative Schools in the city.

Schools have a crucial role in attaining clean water and air and sufficient food and energy for society. The next generation needs information about alternatives to dirty rivers, ozone alerts, and concrete prairies. They must be taught that broken bikes and old tires cannot be gotten rid of, but will hamper community progress unless they are recycled to their proper place in the ecosystem. Schools on all levels have the responsibility to train students in the basic principles of ecology and environmental management that support their very existence.

Hanson Park School, on Chicago's northwest side, is the city's first solar retaliation against spiraling energy costs for education. The existing school was adapted to solar heating and connected to a new, solar-heated structure designed for two hundred handicapped students. Projected costs, allowing for estimated increases in fossil fuels, show an operating savings of $90,000 yearly with solar heating, and a payback period of seven years for the added solar and insulation costs.

THE WOODRIDGE SCHOOL

Environmental education is a "multidisciplinary approach to the study of man's relationship to his natural and man-made surroundings," states the *Environmental Education Handbook for Teachers,* published by the Illinois Office of Education in 1976. This is the focus of a proposal for a completely new kind of school outside Woodridge, Illinois . Chosen for its suitable topography, this site on DuPage County's eastern border lends itself to student interaction with the forces and principles of nature.[10]

Graduates of the proposed Woodridge Environmental School would embody the consciousness that a resource cycled is a resource multiplied. Textbooks, teaching stations, and readily visible processes could inculcate concepts and methods for relating healthfully to the air, water, and soil. As students learn the advantages of resource management, they would form new life-styles and influence others to manage and multiply our natural wealth.

Recognizing the sun as the primary source of energy, the Woodridge model school would harvest this power for heating, cooking, and illumination. It conserves energy by the location and number of windows and doors, by the protective windbreak of trees and mounded earth, and by adjustable thermal shutters and drapes. Access panels to operating equipment and monitoring instruments would enable students to tune in daily to the ecosystem that shapes their physical well-being.

Water, the "vehicle of life," would travel steadily around the land-sculptured Woodridge site. Rainwater and wastewater would flow to detention basins, over a meadow and into a marsh, thereby yielding pollutants and organic nutrients to grass and soil before returning filtered water to underground drains and the water table beneath. The controlled flow would recycle water for healthful reuse by plant, animal, and human life.

As inadequate management of stormwater is a major problem in urban America, the model school would illustrate remedies in a significant way. The EPA-funded Black Creek Project attributes 50 percent of the nation's water pollution to such nonpoint sources as grit, oil, animal debris, air pollution fallout, soil, and fertilizer deposited in streams from paved areas, farmlands, and the atmosphere. Conventional urban developers complicate the problem by forsaking the natural ground filter and channeling pollutants to watercourses that should support aquatic life, enhance recreation, or supply drinking water. This disruption of the water cycle imperils health as well as damaging property by flooding.

Rainwater at the Woodridge School would siphon through openings of the paved road and parking areas to a gravel underlayer and then into the ground. Roof runoff would drain to a sodded basin for detention and for filtering of pollutants. The finished site would discharge no more water to the nearby creek than did the undeveloped plot.

A wastewater reclamation project at the corner of the schoolgrounds would show students the tremendous capacity of soil to change sewage pollutants to plant nutrients. First step in the process is a mechanically aerated pond where oxidation and bacteria stabilizes the organic materials. Next the sewage flows down a sloped meadow, nourishing a crop of thick grass and undergoing further chemical change. Last, a low-lying marsh receives the renovated sewage and extracts further nutrients for abundant plant and animal life. Within a one-third acre space, students witness the vitalizing effects of water and nutrients in building the complex food chain for plants, animals, and humans.

Solid waste management can also be demonstrated at the school, not with a sanitary landfill, which requires a large area, but with a reclamation center for collecting recyclable glass, paper, and some kinds of metal products. Disposal of containers and other trash is a formidable problem in our throw-away economy. At stake are conservation of resources, protection of the environment, and habitable living space. (See Chapter 7 for ways individuals can help.)

Financial returns are modest in a reclamation project, but principles of conservation are taught and both raw materials and energy are saved by recycling certain kinds of trash. It is another case of turning negative environmental factors into positive ones. The project also builds community cooperation and planning for the good of all.

Country windmills used to pump water, grind grain, and saw wood. Today an aero-dynamically styled windmill can spin an electric generator and provide energy to store in a battery for slow-wind periods. At Woodridge a demonstration windmill-power plant would perch on a bluff overlooking the DuPage River floodplain. Wind velocity varies from day to day and month to month, but the power-storage facility would permit the generation of an estimated 10,300 kilowatt-hours of electricity over a year, an amount used by a typical single-family residence. The output could energize several low-power operations at the school for demonstration purposes.

As energy costs rise in the coming years, designers and builders—as well as homeowners and tax payers—will be increasingly conscious of economy, efficiency, and pollution control in the use of power. A new generation needs to learn

105

that nature is a highly profitable partner in producing energy. It is hoped that the Woodridge prototype will be duplicated in several areas throughout the country.

BUSINESS IN THE PARK

Big-city entrepreneurs can secure a new lease on natural living in the suburban office park being built on Chicago's western edge for Trammell Crow Company, international developer and manager of business properties.[11] A luxury hotel and low-to-high-rise office and commercial buildings will overlook an expanse of grass, trees, and man-made lakes in the Itasca Center, a 274-acre site adjoining Interstate 90.

Unique among office parks will be Itasca Center's environmentally intergrated landscapes and commercial operations. All the business centers are readily accessible by car, but core areas are traversed only by attractive walkways. Energy-conserving buildings will exchange heat seasonally with the lake water by heat pumps, and solar technology equipment will be added when it becomes cost effective. A complete water-recycling system from well to recharged aquifer assures an abundance of clean water and absence of flood water in the resplendant setting.

Precisely graded lawns, water-collection swales, gravel-filtering layers, underground drains, and chlorination will collaborate to channel, filter, and recondition both storm runoff and wastewater for return to the lakes or the water table. Itasca Center features the land treatment method of water renewal working its multiple benefits near a metropolitan area and pointing the way toward self-sufficient urban modules of the future.

TAMING THE TORRENT

Most urbanites have a love-hate relationship with water—there's usually too much or too little at the wrong time or place. We cherish pure drinking water but grimace at chemically-rank tap water; we revel in the sight of a distant lake and shrink from contact with its slimy surface; we alternately endure restricted water days and flooded basements; we

Columbia, Tennessee, victims stare at their Federally-insured homes, built in an area which TVA noted was an "easily identifiable" flood hazard area.

marvel at Houston's astonishing growth and gasp at nearby Baytown's sinking buildings that are undermined by excessive pumping of water from the ground. Water is not only a problem today, but an elemental giant that threatens havoc upon enterpreneurs who ignore it in their planning for tomorrow.

This giant can be domesticated by resource planners who know its character and strengths. On small tracts and vast expanses, storm water is being tamed for multiple services. Captured and channeled, it delights the eye, moderates the climate, nourishes vegetation, shelters aquatic creatures, aids industry, replenishes the local supply, and stores surplus volume. Every type of water problem can be turned into an advantage when land developers respect the full range of water's activities.[12]

In Bensenville, Illinois, a new factory was built on a frequently flooded site after bulldozers scooped out lagoons for water retention and elevated the plant site with the excavated dirt. This environmentally planned construction alleviated a downstream flooding problem rather than aggravating it.

In Denver, rooftop ponding eases stormwater run-off problems at the large Skyline Urban Renewal Project. By detaining rain on the roofs and releasing it slowly to storm sewers, the downtown housing project reduces water damage both in its own area and in downstream sections of the city.

A carefully designed tennis court serves as a temporary detention basin for stormwater at an apartment complex in Arlington Heights, Illinois. Knowing that Weller Creek was already overtaxed in peak runoff periods, architects created the double-duty tennis court that captures heavy overflow for a period and resumes its recreational character after discharge of the water and a hosing down of the asphalt surface.

Each of these projects was designed specifically for stormwater problems in the area. The rewards were enhanced community enjoyment and varied economic benefits.

Land developers and zoning commissioners tend to disregard a primary environmental fact: floodplains are valuable land resources. This does not mean that some low-lying marshland should never be improved for human uses, but it does deny the real estate maxim that developed land is the only worthwhile land.

Floodplains exist by necessity, retaining water where it will not damage developed property. Trying to substitute stormsewers for floodplains has turned many residential basements into walled swamps. Floodplains are valuable as water rechargers to filter and replenish the local water supply. They also provide open space for structure-bound urbanites, natural havens for birds and small animals, and a prime laboratory for the study of wildlife, plants, and soil organisms. These multiple values were recognized in the recently issued Presidential Executive Order on Floodplains.[13]

Wheaton, Illinois, typifies many communities' search for environmental and economic balance in land use. Its 140-acre Lincoln Street Marsh witnessed steady encroachment and change as housing and business developments pressed closer and altered the waterflow across its boundaries. Nature lovers valued its spacious stretches sufficiently to preserve a biking and walking trail on the edge of the marsh, but a major test came with a civic commission recommendation to convert the marsh to a golf course.

Aside from the question of suitable disposition of rainwater and the practicality of refashioning a one-time swamp into a dry greensward for golfing, an overriding concern remained. What were the comparative values to the community of an ancient natural preserve where air, water, and soil interacted with plants and animals, as opposed to a picturesque recreational site for serving a minority of residents? Conservationists have prevailed in the early debates of the issue. Some of the property has been acquired and negotiations are underway to preserve the entire site as an educational nature preserve.

As new houses and business complexes swallow urban space, the urgency grows to integrate water management strategy with overall regional planning. Streams and rivers link communities in a watery lifeline that may be constricted or expanded with far-reaching consequences. Though everyone knows water runs downhill, most communities are surprised by sudden floods. The cause is a lack of coordinated planning all along a watercourse and throughout a drainage basin.

The Regional Planning Commission and Forest Preserve Board of DuPage County, Illinois, united to head off water problems connected to Salt Creek, a tributary of the Des Plaines River. As a heavily populated area adjoining metropolitan Cook County, DuPage still has several open areas which can vitally affect environmental conditions in the well-drained river and lake region. Officials of the cooperating agencies agreed that acquisition of a strategic network of floodplains along the main drainage systems and other open lands would conserve natural resources and upgrade living standards in the county.[14]

Guidelines for land acquisition were: groundwater recharge capacity; watershed area for flood prevention; flood-prone sections; surface storage qualities; water-oriented recreation potential. Management of these areas would entail preservation of the wildlife and ecological and aesthetic values. Ideally, the amount and condition of these floodplains and wetlands would guide the type of urban development in the county. After several years in operation, the master plan is proving its worth for fast-growing DuPage County.

Urbanites may question the expenditure of tax funds to set

aside open spaces that could produce tax revenues if developed by commercial and real estate managers. But such developments require increased educational centers, sewage and water facilities, police and fire and court services, and street upkeep—shrinking the net tax benefit substantially and sometimes adding a tax burden. Environmentally, the overtaxing of land, water, and air resources in urban areas precipitates bankruptcy of our life-support systems. Communities that desire enduring prosperity will act to guarantee the perpetuation of their natural reserves of rich soil, sparkling water, and fresh air.

SEVEN
LIFE-STYLES
THAT COUNT

The former president of a theological seminary in "mile-high Denver" whimsically noted the main dilemmas of high-living Americans recently. Said Vernon Grounds, "As a middle-class American, I have habitually eaten too much. Blessed with an abundance, we in the United States face three pressing problems: Who's going to win the game? Where do I park my car? And how do I lose a few pounds?"

Can affluent Americans shut out of their minds the fact that over a billion of the world's people are struggling to survive? Bad housing, insufficient food, and inadequate medical care blight these lives. Tragically, the great majority of the sufferers are children. They lack the strength and resourcefulness to secure their share of food and medicine.

Overeating, overstress, and overpreoccupation with material possessions are by-products of modern technology. While material rewards are legitimate returns for enterprising effort, overindulgent living has always fostered flabbiness of both body and soul.

Naturalist Henry David Thoreau declared: "Most of the luxuries and many of the so-called comforts of life are not only not indispensable, but positive hindrances to the elevation of mankind." Churchman John Henry Newman, glimpsing the same danger in the nineteenth century, cautioned: "A smooth and easy life, an uninterrupted enjoyment of the goods of Providence, full meals, soft raiment, full-furnished homes, the pleasures of sense, the feeling of security, consciousness of wealth—these and the like, if we are not careful, choke up all the avenues of the soul through which the light and breath of heaven come to us."

Few other nations possess such rich natural and human resources as the United States. These include our political freedom and technological capacity. They have enabled our 6 percent of the world's population to use 35 percent of the world's energy each year. But more Americans than ever before are enjoying their abundance less. Many young people—and some older ones—are reacting against the desensitizing effects of overindulgence and the thoughtless damage to Earth's ecosystems.

An international and intergenerational obligation for contemporary Americans was pointed out by energy specialist S. David Freeman. "To the environmental and political constraints on our growth in energy production should be added a moral constraint. If all the people on Earth are eventually to enjoy an adequate level of material well-being, we will need to adopt a new ethic which regards waste as a form of theft. For if we continue a self-indulgent, disposable society where the cycle continually is to dig, burn, build, and then discard, we are stealing from our children and grandchildren the planet's resources.[1]

PERSONAL LIFE-STYLES

Though far-ranging improvement of the environment will require definite governmental guidelines and broad industrial cooperation, the attitudes and actions of individuals also shape the environment and social practices significantly. Public opinion and involvement have turned the nation around at pivotal points in its history.

In an era of scarcity, colonial New Englanders were said to exhort their children: "Use it up, wear it out, make it do, or do without." That is a far cry from the reported philosophy of a current auto-making executive: Planned obsolescence is another word for progress. This decadence by design in the auto industry amassed unparalleled profits, to be sure, but it transmuted personal wastefulness into a virtue. The slightly

112

old became slightly disreputable, the conspicuously new became fashionable and prestigious. Today the widespread waste of energy, resources, and human talent in boosting the Gross National Product is a notorious American habit.

Some conservationists insist that a drastic reduction in living standards is necessary to avoid ecological breakdown. Simplified life-styles will indeed gain numerous benefits for individuals and society, but the real key to restoring the environment and justly distributing the Earth's natural wealth appears to be a more efficient use of energy and resources. In every land and culture, the modern maxim might well be: Waste not, want not. Waste and inefficiency are luxuries that even the richest nations cannot afford.

Efficiency can be a complicated matter in highly industrialized societies. It requires knowledge of interacting ecological systems and the comparison of one advantage against others. It depends for success on responsible and largely voluntary choices by producers and consumers, because the majority of efficiency choices are too numerous and personal to enforce by law. Each person's "quality of life" involves intangible tastes and individualized needs that cannot be legislated. Like morality, efficiency must be practiced by individuals in order for society to survive.

War against waste in a throwaway society is necessarily a relentless campaign. We need to neutralize Madison Avenue's bombardment of consumers by discerning examination of glamor-gilded advertisements. We need to close our checkbooks and shut down regiments of energy-voracious machines, at least until efficiency ratings are proved and approved. We need to inspect carefully those legions of products peddled before us daily before we welcome them into our homes.

Coasting along in the consumerism tide seems so much easier. In the past, technology lowered the cost of new products and raised the earnings of workers. It reduced toil and increased leisure. Comfortable homes became mini-castles, and lords and ladies of suburban manors expected to live happily ever after. But somewhere along the technological trail many travelers discovered they were not enjoying "the good life."

Surrounded by good things, we tended to scorn common

THE FINAL THROWAWAY

things. Our struggles to outwit nature turned into contests to outshine neighbors. Prospering businesses went to three shifts and seven-day work weeks, and frenetic family members went in all directions—separately.

At home, new gadgets seemed very important and yet highly expendable. When rivet or power cord broke on an appliance, a repair man was hard to find. It seemed more efficient to toss the defective merchandise and buy a new one. Some kids grew up thinking "fix it" was a punishment. Time became the high-priced commodity—time to make more things, more money, and more plans.

And here we are: masters of Earth's treasures and slaves to our machines. People have been dethroned by their products.

But individuals here and there have identified a void in the midst of plenty. It's the absence of humaneness, of enduring meaning, of relatedness to our wondrous habitat. Columbia University sociologist Amitai Etzioni observed: "In the last ten to fifteen years, the principle of working hard to consume hard has been challenged from the inside. One-half of the population no longer buys one-half or all of the equation."[2] Today there's a quest for compassion with our comfort, conscience with our convenience, and creativity with our consolidation. This new life-style accents responsible choices.

COUNTING COSTS

People with environmentally sensitive life-styles will discern products and advertising that flout stewardship of resources. The plastics revolution offers a case in point. Glass made from sand, an abundant natural resource, is a better all-around choice for a container than is plastic. Glass is reusable and easily recycled, and it is not dependent upon petroleum for its manufacture, as is plastic.

Returnable versus discardable containers for beverages has become a controversial matter in many states. Here are benefits cited in the proposed Illinois legislation: (1) saves energy and raw materials, (2) saves tax assessments for litter pickup and waste disposal, (3) reduces manufacturers' and hence potentially the consumers' costs, (4) safeguards children from pull tabs and thin-glass throwaway bottles, (5) adds jobs for unskilled workers who have high unemployment rates, and (6) makes the environment more pleasing aesthetically.

Clear-cut decisions that benefit the environment are not always obvious. In comparing the trade-offs of glass versus cardboard cartons for milk, for example, pleas to "use glass and save a tree" may not be the best choice. In this instance, cardboard is degradable, whereas the energy and washing materials needed for returnable glass bottles may take a greater environmental toll. Fast-growing pulpwood trees such as aspens or willows may be successfully managed and economically supplied to the cardboard industry. Adequate information for making the best decisions is lacking in many production areas, but scientific diligence can secure it. Answers will change as news methods and materials alter the bottom line of net energy and environmental impact.

The Reynolds Aluminum Company, among others, responded to environmental and efficiency concerns in paying $53 million for 7 billion recycled aluminum cans in 1978. Recycling glass, paper, and steel also conserves materials and energy. More than two hundred U.S communities have trash recycling projects, with about fifty of them compulsory.

Personal adjustments in energy use for transportation and home heating and cooling also can improve the energy picture. Walking and bicycling for physical fitness and fuel savings is a promising trend. Smaller cars are here to stay, with signs that rising gas prices will press car manufacturers into innovative efficiencies that will also diminish pollution of the atmosphere.

Though solar energy is the obvious long-term solution to energy needs, this option is already being implemented in many areas of the country. Pushing the wave of the future, the Department of Energy has completed an installation in New Mexico where an entire community is powered by photovoltaic conversion of solar energy into electricity for lighting homes, powering a community water pump, and running a series of refrigerators. This and other demonstration installations now require heavy federal subsidies, but they have the potential to stimulate investors and manufacturers to promote innovations that can accelerate a swing to solar with consequent reduced costs.

HOME FRONT GAINS

Practicing what he preaches in the classroom, coauthor Ray Brand has completed two parts of a three-phase plan to bring more of the sun's rays into his life-style. Phase one involved the installation of a water-to-air heat pump and 1,000-gallon water storage tank. On summer days the pump cools the house by transferring warm inside air to the water in the tank. At night the hot water is pumped outside to a heat exchanger near the ground where it cools and is returned to the tank for the next day's cycle.

In winter a fireplace grating with water-filled pipes (phase two) sends heat from the evening fire to the storage tank. When the thermostat calls for more BTUs, the heat pump draws on water-stored heat before the gas furnace is used.

Upon completion of phase three, which involves solar collecting panels on the house's south-facing roof, the gas burner will trip on even less frequently. The economics of the project look better every time oil-producing countries meet for another price-boosting conference. A tax credit of 30 percent of the first $2,000 plus 20 percent of the next $8,000 for solar equipment expenditures and of 15 percent of the first $2,000 for insulation materials is another encouragement.

When day-to-day habits are shaped by personal choices to save energy and raw materials, we confirm our commitment to efficiency and environmental vitality.

President Carter was disappointed at the public response to his appeal for "the moral equivalent of war" against advancing energy shortages. He told a convention of the American Society of Newspaper Editors in April 1978: "We want something to be done about our problems—except when the solutions affect us. We want to conserve energy, but not change wasteful habits. We favor sacrifice, as long as others go first."

Conservation seems unpopular—even unpatriotic—in a waste-creating culture. Yet when a counter-culture vanguard demonstrates the advantages of efficiency, discerning Americans will recognize that wasteful living has depressed, not elevated, the quality of life. A war on waste will not be won by a few altruistic commanders, but by a mobilized citizenry amassing countless victories in daily efficiency.

Practical suggestions for avoiding waste can regularly be

found in newspapers, magazines, and books. Despite a few premature directives that turned out to be in error, *Everyman's Guide to Ecological Living*, by Cailliet, Setzer, and Love, is still one of the broadest and soundest "What can I do?" compilations.[3] Another source of ideas is *The Environmental Handbook*, sponsored by the Friends of the Earth organization.[4]

Proposed life-styles in various publications range from spartan renunciation of modern facilities to revolution-by-legislation campaigns. Since circumstances and priorities differ for each person, it seems prudent to base life-styles on personal choices that reflect efficient use of available resources. No single life-style can be effective or satisfying for all. As individuals discover the economic, health, and moral benefits of ecological living, many will make efficient methods and materials their preferred choices.

WORLD POPULATION GROWTH

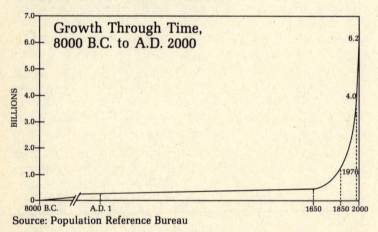

Source: Population Reference Bureau

THE PEOPLE PROBLEM

Probing deeper into personal life-style patterns, we encounter the sensitive topic of family size. Since the beginning of the environmental movement, the population "explosion" has been a serious concern to people alarmed over current widespread starvation.

118

G. Tyler Miller, Jr., illustrated the rapid increase in population this way: "Suppose you decided to take only one second to say hello to each new passenger added [to Earth] during the past year. Working twenty-four hours a day, it would take you two and one-half years to greet them, and during that time 183 million more persons would have arrived, putting you almost six and one-half years behind in your greeting program."[5]

Significantly, however, the average rate of increase in world fertility dropped from 4.6 to 4.1 births per woman between 1968 and 1975. And in 1976 the first worldwide decrease in growth rate was recorded: from 2 percent growth in 1966 to 1.9 percent in 1976.[6] Without citing causes, the U.S. Census Bureau's demographic expert, Samuel Baum, commented: "The changes are small, but they are very significant. This is the beginning of a trend, and it's happening a decade earlier than expected."

Though rising abortion rates and governmental family-planning programs have affected population growth, a rising standard of living in country after country has depressed the birth rate by voluntary limitation. Upgrading living conditions to stabilize population by free choice is highly preferable to government sanctions for controlling family size. Information programs on family planning contribute to the socio-economic improvement.

Numerous authorities assert that worldwide malnutrition and starvation are not due to incapacity to produce sufficient food, but to the indifference of economic and political leaders. The problem is not too many people, but too few caring people. Again, quality instead of quantity makes the difference.

Churchgoers who have demonstrated concern for the needs of the world have a high stake in the success of a national energy program. Unless efficient practices bring pyramiding energy costs into line with personal incomes, contributions for Christian charities and missions will shrink sharply.

Looking forward to an International Consultation on Simple Life-Style sponsored by the World Evangelical Fellowship in early 1979, Anglican clergyman John Stott declared:

"We may not all give an identical definition of justice and injustice, or share the same economic theories and remedies, or believe that God's will is an egalitarian society in which even the slightest differences of income and possessions are not tolerated. But we are all appalled by poverty, that is, by the immense numbers of people who do not have enough to eat, whose shelter and clothing are woefully inadequate, and whose opportunities for education, employment, and medical care are minimal. Every sensitive Christian should be shocked by this situation and never grow so accustomed to it as to be unmoved by it.

YOUNG LIFE

Change often comes slowly in a democratic society, but frequently the younger generation leads the way. Such was the case in the early days of the environmental movement when Earth Day, April 22, 1970 was promoted and celebrated by student groups throughout the country. Conferences, rallies, protest marches, T-shirts, buttons, posters, and environmental newsletters dotted university and college campuses and spread to high schools and environmental units in elementary schools.

Shifting from the campus violence of the 1960s, students formed groups concerned with improving the environment. Existing conservation organizations took up the cue and became more aggressive in opposing pollution and supporting environmental legislation. The National Audubon Society and its local chapters continued to sponsor bird hikes and an annual Christmas census of bird species, but they began to question rampant dam construction and channelization of rivers across the nation. Professional lobbying groups such as the Environmental Defense Fund became a force to be reckoned with by those who tended to slight new environmental legislation.

Another group founded in 1969 by David Brower, the Friends Of the Earth (FOE) organization, became a leading spokesman for environmentalists. Basic to the influence of FOE was the timely publication of *The Environmental Handbook* in 1970 for Earth Day and the environmental teach-ins. It went through three printings in three months.

Citizen advocate groups, spawned by the early successes of

"Ralph Nader's Raiders," sprang up in various locations and promoted environmental concerns along with other causes. In the Chicago area, Citizens for a Better Environment struggled to survive on voluntary contributions of time and money. Then, as a result of court settlements in their favor plus a widened base of support from Chicago suburbs, this organization expanded to branch offices in Washington, D.C. and San Francisco. Today they publish a monthly news brochure, *The CBE Environmental Review*, and employ research analysts with credentials in the natural sciences as well as lawyers trained in environmental law.

Courses on the environment sprouted across the country. On many campuses "Man and the Environment" courses became an integral part of the curriculum. Textbooks and popular paperbacks streamed into bookstores on every aspect of the environmental movement. Groups outside of class formed to take up pragmatic aspects of action-involvement projects.

One such group at Wheaton College, called "Wheaton Students for Environmental Responsibility," holds regular meetings, sponsors guest speakers, organizes campus projects such as water conservation contests and energy utilization studies of campus buildings, and presents an annual chapel program to integrate Christian faith with environmental issues.

Many informed students are committed to environmental objectives, as shown by the following quotes taken from a recent church class for college students:

"Christians should realize we do not own the earth and it is not ours to do with as we see fit."

"Nothing but our own lack of concern prevents us from effective pollution control and proper management of our resources."

"I came to understand that 'the earth is the Lord's and the fulness thereof.'"

"I learned that God entrusted his world to man to enjoy and delight in and to manage."

"I realize that I am personally accountable to God for my actions and negligence in this area."

Other statements from class members referred to involvement by voting, writing to congressmen, and participating in community activities to improve the environment.

These students perceived that their efforts can make a difference, that the tide of environmental decay may be turned, and that they can be part of a hopeful future.

Environmental economist E. F. Schumacher spotlighted the innate rights of the world's needy people in his book *Small Is Beautiful*. Echoing the words of Jesus, he addressed both multinational corporation leaders and average "first world" citizens when he wrote: "Everyone to whom much is given, of him will much be required."[7] It is clear now that personal investment in efficient living styles will yield rich dividends at home and around planet Earth.

EIGHT
THE POLITICS
OF ECOLOGY

Change to efficient life-styles by individuals and families is essential for building a new popular consensus and setting a new national direction in environmental care. The waste-not, want-not standard must extend to government policies, however, to assure that vast public funds support efficiency rather than wastefulness. This requires the informed participation of citizens at local, state, and national levels.

World food production and equitable distribution, as formidable a human problem as any we face, can be alleviated by the united action of Americans.

WORLD FOOD

As Frances Lappé and Joseph Collins show in their book, *Food First: Beyond the Myth of Scarcity,* current world grain production could provide every person on earth with more than 3,000 calories a day. So total available food is not the problem. One-third of the world's grain and about one-fourth of the fish catch are fed to livestock, yet neither is that the crux of the problem.[1]

Research by the *Food First* team revealed that food production kept pace with or exceeded the rate of population growth among 86 percent of the people living in developing countries. But in places such as El Salvador, Senegal, and many other countries, major landowners planted cash crops and sold the harvests to multi-national corporations—many of them American—while native laborers went hungry.

So—what can American citizens do about the economic imperialism of giant corporations and the callousness of foreign land merchants whose act may be unethical, but are thoroughly legal?

We can persuade our congressional representatives that we will no longer permit our tax money to support already prosperous farmers instead of the needy, nor have these aid funds channeled to countries to gain political or military advantage rather than to feed the hungry.

Present U.S. assistance programs provide grants to overseas land owners who have the knowhow and property to capitalize on the assistance. This assists only a minority, and they do not share our gifts with their landless, hungry neighbors.

U.S. legislation also allows the donation of food, credit, and equipment to nations that will serve our foreign policy objectives under the "aid" program. While important, those goals should not be pursued under the pretext of humane sharing. As now formed, U.S. programs of assistance to needy nations are far from generous, efficient, or realistic. In a United Nations listing, the U.S. ranks 13th among 16 nations in providing aid to nations needing food.

In the fall of 1978, a national television audience gained new insight as a result of citizen participation in the 25,000 member Bread For the World (BFW) organization. A series of documentary programs called "The Fight for Food" explored countries where nutritional problems exist and progress has been made in meeting this basic human need. Concluding the series was a lively forum involving leaders from both "have" and "have-not" nations of the world.

Dovetailing with this presentation was a BFW series of workshops throughout the United States on food and hunger. Many attendants and members of this organization then sponsored discussion groups in their communities. These encouraged participation in formulating national policies for creating a global grain reserve. Later Congress enacted a measure based on BFW guidelines that initiated a grain reserve program.

MENDING HOME FENCES

Not all the important decisions are made in Washington, D.C. County and state officials and laws also shape the local quality of life in many ways. To illustrate, DuPage County, Illinois, had a 1979 budget of $103 million. That's enough to deserve close attention from its citizens. Will the county funds go into priority projects? Do the citizens know the goals of their elected officials? If not, democracy is failing at the county level, and a non-functioning democracy may in time fail the people on all levels.

County organizations are strategic in many environmental issues since they facilitate regional planning that touches many cities. As special-interest groups have organized to advance their causes through county government—the basic building block in national political parties—the larger public must not lose its prerogatives through neglect of this influential administrative arm.

Environmentally minded leaders in educational and political circles have put the states of Wisconsin and Oregon in the forefront of ecological progress. In Illinois, the attorney-general's office, has taken giant industrial polluters to court and emerged with convictions. One recent judgment led to the addition of major antipollution controls by United States Steel to protect Lake Michigan from further deterioration. The plant is in Gary, Indiana, between Illinois and Michigan border-shorelines, so citizens from all three states participated in the legal action and benefited from its outcome.

The State of Illinois also took on the federal government in a toxic-chemical case. The suit charges the federal EPA with approving a site near Wilsonville as meeting regulations for safe dumping of radioactive wastes, though such regulations had never been established. State environmentalists are particularly concerned because Illinois has become a nationwide storage center for radioactive wastes. So far, permanent disposal facilities for this near-permanent hazard exist only in the untested plans of federal agencies.

Diligent environmentalists caught the federal EPA with grimy hands in 1977. While identifying 600 industries and 2,000 cities as missing the statutory deadline for constructing adequate sewage facilities, the federal EPA kept silent about hundreds of federal facilities continuing to pollute water-courses with their sewage. The Chanute Air Force Base in Rantoul, Illinois, was operating a sewage treatment plant designed for 5,000 people which became overloaded with three times that number. When the information leaked out, the EPA vowed to clean up its own "house."

The five Great Lakes are the most prominent natural feature of the Midwest environment. To monitor and help manage these repositories of about 20 percent of the world's fresh water supply, the Great Lakes Basin Commission attempts to coordinate planning for the land and water resources of the area. Eight states and twelve federal or interstate agencies are represented on the Commission, and along with 3,000 agencies—from township boards to the U.S. Congress—they are potential partners in this gigantic environmental collaboration.[2]

At the state level, a newsletter of the Illinois Environmental Council keeps concerned residents informed about environmental activities in the state capital. Council goals are: protect Illinois citizens from hazardous materials, including radioactive wastes; promote the use of returnable beverage containers instead of throwaways; protect Illinois land from the ravages of strip mining; resist "urban sprawl" and support more appropriate uses of land; demonstrate that environmental responsibility need not threaten the economy or cause job losses; uphold the federal clean air and water legislation; develop a system of bicycling and hiking trails on abandoned railroad corridors; promote the use of solar energy and energy conservation; preserve scenic and natural areas, including rivers, lake shorelines, forests, and prairies; and protect our heritage of historic areas and archaeological sites.[3]

THE WASHINGTON ARENA

Citizen involvement in environmental issues is crucial at every level. The history of national water quality legislation records a long-term struggle, as indicated in a 1972 report. "The first specific, comprehensive federal thrust toward water pollution control was embodied in the Act of June 30, 1948."[4]

Details of this early law provided low-interest loans up to $250,000 for local communities to construct sewage and waste treatment works. Funds for all purposes covered by the law amounted to $24.3 million a year, and the Act of July 17, 1952 continued the terms of the legislation through fiscal 1956. Other key laws followed in the National Water Pollution Control Act of 1956, the Water Quality Act of 1965, and the Clean Water Restoration Act of 1966. The latter authorized federal expenditures of about $3.5 billion for fiscal years 1967 to 1971, although actual appropriations came to about one-half that amount. The magnitude of the growing problem is reflected in the increase of approved funding from $24 million to $710 million per year. Yet the nation's water continued to get dirtier.

The Muskegon land treatment success gave new impetus and direction to the government's clean-water campaign in 1972. Congress passed the Federal Water Pollution Control Act Amendments which advocated wastewater recycling for the nation's cities. But the recommendation fell far short of legal requirement, and communities continued to pollute their waterways. It was again time for citizen action.[5]

A lobbying group known as the Citizens for Clean Water Committee got busy in Washington, D.C. and helped to bring through Congress a new law with teeth in it, the Clean Water Act of 1977 (Public Law 95-217). Pertinent to the environmental perspectives of this book are these selected quotes from the law (which covers forty-four pages of government print):

> The Administrator shall not make grants—to any State, municipality, or intermunicipal or interstate agency for the erection, building, acquisition, alteration, remodeling, improvement, or extension of treatment works unless the grant applicant has satisfactorily demonstrated to the Administrator that innovative and alternative wastewater treatment processes and techniques . . . have been fully studied and evaluated by the applicant, taking into account the more efficient use of energy and resources.

> Not less than one-half of one per centum of funds alloted to a State—shall be expended—for treatment works utilizing innovative processes and techniques.

> The Administrator shall submit to the Congress by

October 1, 1978, a report on the status of the use of municipal secondary effluent and sludge for agricultural and other purposes that utilize the nutrient value of treated wastewater effluent.[6]

Funding for this wastewater statute dwarfs earlier allocations. Authorizations range from 1 billion to 4.5 billion dollars per fiscal year from 1977 through 1982, with the total maximum expenditures not to exceed $30 billion for the six years. Citizen taxpayers, beware: environmental issues and costs affect more than ecologists and sanitary engineers. Fiscal and environmental solvency may ride on the recycling extent of this national program. It is vital that other environmental and energy-related policies regarding air pollution, solid waste mangement, land use planning, and preservation of wilderness are formulated with similar built-in efficiency.

Public programs and private business are often regarded as antagonistic to each other. That needn't be so. As public planning reaches farther into business operations, and as industrial activity affects the environment more widely, commercial goals should merge with society's. Monetary profit and resource efficiency have become closely related in both the private and public sectors. Shared progress, rather than punitive regulations, should be the mutual philosophy of planning for the country.

According to the Bible, "Righteousness exalteth a nation" (Proverbs 14:34), meaning that personal and public integrity elevate a nation—or a state or city. Such a nation manages its resources so that it constructively influences the far corners of the world.

GOING GLOBAL

Since its beginning in 1945, the United Nations has attempted to probe global environmental problems. The United Nations Environmental Program (UNEP) sponsored conferences on deterioration of the ozone layer, environmental education, serious water shortages and poor water quality, and the effect of human activities on the expansion of the desert. Such conferences have set up committees to submit policy proposals to the General Assembly.

In the management of ocean resources and the control of ocean pollution, the Third U.N. Conference on the Law of the Sea, begun in 1973, obtained these preliminary agreements by 150 nations.

> Territorial seas shall extend no farther than 12 miles offshore;
> Coastal nations' economic zones may extend as far as 200 miles offshore;
> And global cooperation is required for the conservation of highly migratory species of fish.[7]

Although additional refinements are needed, these first steps move in the right direction. Ocean productivity is largely restricted to estuaries and coastal zones because the open seas, comprising 90 percent of marine waters, are the equivalent of a biological desert as far as productivity is concerned.

To implement the economic zone agreement, the U.S Congress passed The Fishery Conservation and Management Act of 1976, and President Carter promptly used it to prohibit commercial whaling within the 200-mile zone off U.S. coasts. This action reinforced the program of the sixteen-nation International Whaling Commission that seeks to prevent extinction of remaining whale species.

Several conventions have gained ratification of environmental agreements. One of these concerning conservation of migratory birds and their environment was signed by the United States and the Soviet Union in November, 1976. Earlier the same year twelve nations signed an agreement in Barcelona, Spain, "to prevent, abate, and control pollution of the Mediterranean and to protect and enhance the marine environment in that area."[8] This document stressed the importance of preventing fuel dumping from planes and ships as well as pollution run-off from lands around the Mediterranean.

The International Commission on Atomic Energy has made little progress in obtaining commitments from every nation in "the atomic club." Atmospheric testing continues, and though the noise may not be heard around the world, the radioactive fallout is likely to be widely felt in coming years. Research and widespread publicity concerning nuclear hazards are the best means for restraining atomic proliferation both in this country and abroad. Nuclear reactors and nuclear war are unprecedented threats to human welfare. International agreements are needed to prevent a future nuclear holocaust.

Maurice F. Strong, director of UNEP in 1972, summed up his global imperative for the environment in this way:

1. Populations must be stabilized.
2. There must be a worldwide program to conserve scarce resources.
3. New models for economic and social progress should be elaborated and adopted.
4. There must be a much larger flow of resources between rich and poor countries, with heavy emphasis on providing basic social services to the poorest sectors.
5. Science and technology must be mobilized on a worldwide scale to reduce our ignorance about environment, resources, and population, and to help devise new ways of improving the human condition.
6. The resources of the oceans beyond national jurisdictions must be put under international control to assure that the principal beneficiaries of their exploitation are those living in the developing world.[9]

For individuals and nations, environmental degradation has joined the age-old problems of war, famine, and pestilence as a foe of mankind. As with scientific and economic advances in the past, American knowledge and commitment can blaze environmental trails for the rest of the world.

NINE
THE BLOSSOMING EARTH

Let's sketch an ecologically designed community with the technological changes that would eliminate most environmental pollution and provide a surplus of energy in a model city.

DESIGN FOR LIVING

The planning principles would be:

1. The environment that sustains all life is a single system of interacting materials and forces whose chain-reaction effects are increasingly understood by scientists.

2. The environmental system retains all its elements although they change form and location.

3. Environmental pollutants are often potential resources in the wrong place.

Putting these principles into community practices would evolve programs that:

1. Recycle sewage nutrients onto farmland or open areas instead of discharging them into waterways which become oxygen depleted and biologically sterile; and link auxiliary projects that conserve energy use or enhance living, such as industrial water cooling or recreational space.

2. Reclaim paper, glass, and metals from solid wastes to conserve raw materials and energy used in processing them.

3. Generate methane gas for heating, cooking, and lighting by processing organic wastes in anaerobic digesters.

4. Build houses with passive solar heating-cooling systems, and with active solar systems where cost-efficient.

5. Incorporate stormwater collection lagoons to prevent flood damage, reduce water treatment costs, add beauty, and augment plant and animal ecosystems.

6. Place vegetation barriers along highways and railways for noise suppression and aesthetic values.

7. Construct energy-efficient in-city transport systems with bike trails, public conveyances, and vehicle-pollution standards.

8. Educate all students in basic ecological principles and their practical consequences.

9. Shape land-use and community-expansion policies according to the supply of manageable resources in the area.

10. Establish an "Environmental Police Department" to monitor and publicize local conditions.

The development of such communities through a partnership of technology and ecoscience would restore health, beauty, and economic stability to the residents and their surroundings. Others have envisioned the bright future issuing from this kind of responsible, knowledgeable management of the environment. It has been pictured as follows by the founders of a new environmental group, the American Clean Water Association.

> The time is ripe for a redesign of the American environment in ways harmonious with nature and consistent with the principles of natural science, particularly the emerging science of ecology.
>
> The public wants it. The American love of the great outdoors, combined with the well-publicized warnings of the experts and environmentalists, increased public understanding of environmental issues, and the daily reports of first-hand experiences with foul air, dirty water, sludges, energy shortages, spills and leaks of toxic substances, explosions, outbreaks of cancer and the like, have made environmental improvement an issue

deeply appealing to a broad segment of the American public. The same people who voted in Proposition 13 overwhelmingly approved more than $100 million in water improvement bonds.

The law now requires an environmental redesign. In the past ten years, Congress and the states have passed dozens of laws on air and water pollution, toxic substances control, resource conservation and recovery, flood control, wilderness preservation, rural and urban development, soil conservation, safe drinking water, improved health care, worker safety, housing, and energy. Together, these laws contain the rough sketches of a new America.

They give a hint of what we can hope for. Clean air and water abound. Resources are conserved, recovered, and recycled. The country is rich in open space, wilderness, and recreational opportunities. Our water-management programs harness the natural beneficial properties of soils, flood plains, forests, wetlands, and agricultural lands. Technologies are appropriate in size and character to the environments in which they are placed. Non-structural solutions to environmental problems are encouraged. The toxification of the air, land, water, and ourselves has ended. Workers are freed from the burden of hazardous on-the-job conditions. Adequate supplies of clean energy power an economy that can deliver on the promises of housing, health care, education, and employment. Our life-support and other systems are integrated and coherent. All this is to be accomplished in the most cost-efficient manner possible, with extensive participation by an informed and educated public.

Large amounts of public funds have been dedicated to the achievement of this new world. The tens of billions of tax dollars and the regulatory activities associated with them have created vast new fields of economic activity for architects, planners, equipment manufacturers, industrialists, utility contractors, soil conservationists, municipal and state governments, lawyers, land developers, farmers, economists, bureaucrats, academics, and many others.[1]

133

Is that an impossible dream? History tells many success stories of people united and motivated by a desire to improve their living conditions. "Without a vision, the people perish," declared the biblical prophet. But with faith and a compelling vision, enterprising people move mountains.

Momentum is building across America for economic-environmental collaboration. It is evidenced in the national clean-water campaign, a citizen-legislator drive that gives promise of far-ranging effects on the hydrosphere and the marketplace. It was exciting to see environmentally concerned individuals move Congress toward passage of the Clean Water Act. This legislative triumph should be repeated in the energy field.

MANAGING OUR RESOURCES

Garrett Hardin, known for his "tragedy of the commons" and "lifeboat ethics" concepts, proposes government management of people to overcome resource shortages. He favors good care of the few rather than minimal care of the masses. But those are not the only options. By managing resources instead of programming people, the energy crisis can be eased in a climate of freedom and productivity. The key is efficient use of energy in every area of life.

Massive development of clean, safe, unlimited solar energy need not be postponed for another decade or generation. The experts who say solar harvesting is not yet economically competitive with other power sources are reading a partial balance sheet. They are ignoring many subsidies supplied to current forms of energy, and they are excluding enormous social and material costs connected with pollutant fallout from fossil and nuclear fuels. If all these costs were taken into account, the economics of solar harvesting would improve spectacularly.

In past years, the low immediate cost of fossil fuels often led architects, builders, and engineers to construct an energy-wasteful society. Today our understanding of environmental principles confirms that we can neither afford nor survive the long-term ecological damage caused by profligate energy use. Achieving an ecological revolution in personal habits and national practices will generate widespread economic and social renewal.

Cooperation with ecological realities is rebuilding a vital nation in the Mideast. Blending development efforts with nature's cycles, Israel is causing the desert to bloom with life. In a land denuded of trees and bereft of top soil by centuries of neglect and warfare, forests are again flourishing and arid regions are blossoming under the supervision of water and land resource managers. Vegetables and a hardy new strain of coffee thrive in the sandy Negev region, orchards and vineyards deck terraced hillsides, and surplus harvests go for a handsome price to nations who haven't yet discovered how to mine organic gold from their wilderness.

The bountiful amenities enjoyed by living in harmony with Earth's environment were described in the Jews' ancient scriptural praise to the Creator:

> He makes springs pour water into the ravines;
>   it flows between the mountains.
> They give water to all the beasts of the field. . . .
> The birds of the air nest by the waters;
>   they sing among the branches. . . .
> He makes grass grow for the cattle,
>   and plants for man to cultivate—
>   bringing forth food from the earth
> (Psalms 104:10-14, NIV).

Such a view takes us beyond the simplistic choice frequently offered, either a clean environment or a growing economy. We see that economic and environmental well-being can develop simultaneously when we apply appropriate technology. Environmental quality is the gateway to a flourishing economy.

The energy-pollution crisis will be alleviated and eventually resolved. Every citizen can contribute to this revolution which will strengthen economic foundations while improving the quality of life. By joining the Earth revolution, the life and health you save will be your own today and your children's legacy tomorrow.

NOTES

ONE *What Happened to Utopia?*

1. David Sarnoff, "Preview of the Next 25 Years," *Fortune*, January 1955.
2. M. W. Thring, "A Robot in the House," ed. Nigel Calder, *The World in 1984* (Baltimore: Penguin Books, 1964).
3. Herman Kahn and Anthony J. Wiener, *The Year 2000* (New York: MacMillan Co., 1967).
4. This series of scientific prognoses appears in *Toward the Year 2018*, ed. Foreign Policy Assn. (New York: Cowles Education Corp., 1968).
5. G. Rattray Taylor, *The Biological Time Bomb* (New York: New American Library, 1968).
6. Takashe Makinodan, *Journal of Experimental Medicine*, Vol. 144(5), p. 1204-1213, 1976.
7. Barry Commoner, *The Closing Circle* (New York: Alfred A. Knopf, 1971).
8. *Ibid.*
9. Barbara Ward and Rene Dubos, *Only One Earth* (New York: W. W. Norton and Co., 1972).
10. Dennis L. Meadows, et al., *The Limits to Growth* (New York: Universe Books, 1972).
11. Mihajlo Mesarovic and Eduard Pestel, *Mankind at the Turning Point* (New York: New American Library, 1974).
12. Paul R. Ehrlich, Anne H. Ehrlich, and John P. Holdren, *Human Ecology: Problems and Solutions* (San Francisco: W. H. Freeman and Co., 1973).
13. S. David Freeman, quoted in "The Energy Crisis," *Time*, 7 May 1973.
14. Paul R. Ehrlich, Anne H. Ehrlich, and John P. Holdren, *Ecoscience: Population, Resources, Environment* (San Francisco: W. H. Freeman and Co., 1977).
15. Alvin Toffler, *The Eco-spasm Report* (New York: Bantam Books, 1975).
16. *U.S. News & World Report*, 6 August 1979 (Data from the International Monetary Fund and *Oil and Gas Journal*).
17. S. David Freeman, et al., *A Time to Choose, The Report of the Energy Policy Project of the Ford Foundation*, 1974.
18. Nicholas Wade, "World Food Situation: Pessimism Comes Back into Vogue," *Science*, Vol. 181, April 1973, pp. 634-638.

19. M. F. Franda, "India and the Energy Crunch," *Fieldstaff Reports* (Hanover, N.H., 1974).
20. David Pimentel, et al., "Food Production and the Energy Crisis," *Science*, Vol. 182, November 1973, pp. 443-449.
21. Report, "Our World Gets Smaller," *U.S. News & World Report*, 18 July 1977.
22. Andre Fontaine, "Europe: Mounting Pressure For Change," *Manchester Guardian* (England), 17 April 1977.
23. John M. Kramer, "Environmental Problems in the U.S.S.R.: The Divergence of Theory and Practice," *Journal of Politics*, Vol. 36, November 1974, pp. 886-899.
24. "Draft Environmental Report On Haiti," Science and Technology Division, Library of Congress, January 1979.
25. Jerry Cheslow, "Can Haifa Cope With Pollution?" *Jerusalem Post*, 11 February 1977.
26. Haynes Johnson, "The Lindbergh Legacy," *Manchester Guardian*, 29 May 1977.
27. Lewis Mumford, *Myth of the Machine* (New York: Harcourt, Brace, and World, 1966).
28. Quoted by Andreas de Rhoda, "The Next 200 Years," *Christian Science Monitor*, 28 July 1976.
29. Wilfred Beckerman, *Two Cheers for the Affluent Society* (New York: St. Martin's Press, 1975).
30. Herman Kahn, William Brown, and Leon Martel, *The Next 200 Years: A Scenario for America and the World* (William Morrow & Co., 1976).
31. Donald Moffit, ed., *The Wall Street Journal Views America Tomorrow* (New York: Amacon, 1977).
32. Ernest Schumacher, *Small Is Beautiful* (New York: Harper & Row, 1973).

TWO *Health in the Chemical Cauldron*

1. John Maddox, *The Doomsday Syndrome* (New York: McGraw Hill, 1972).
2. James O. Jackson, "The Garden of Eden Is Being Paved Over," *Chicago Tribune*, 10 April 1977.
3. Lippman, M. and R. B. Schlesinger, *Chemical Contamination in the Human Environment* (New York: Oxford University Press, 1979).
4. I. M. Brodo, "Lichen Growth and Cities: A Study on Long Island, New York, *Bryologist* 69 (4): 427-449.
5. D. L. Hawksworth and F. Rose, "Qualitative Scale for Estimating Sulfur Dioxide Air Pollution in England and Wales Using Epiphytic Lichens," *Nature*, 227:145-148, 1970.
6. Laurent Hodges, *Environmental Pollution* (New York: Holt, Rinehart, and Winston, 1977).
7. Ronald Kotulak, "Farm Crops Stunted By Sulfur Dioxide," *Chicago Tribune*, 3 December 1978.
8. W. A. Sinclair, "Polluted Air: Potent New Selective Force In Forests," *Journal of Forestry*, 67:305:309, 1969.
9. Samuel S. Epstein, *The Politics of Cancer* (San Francisco: Sierra Club Books, 1978).
10. R. J. Bazell, "Lead Poisoning: Zoo Animals May Be the First Victims," *Science*, July 1971.
11. G. L. Waldbott, *Health Effects of Environmental Pollutants*, (St. Louis: C. V. Mosby, 1978).

12. L. B. Lave and E. P. Seskin, "Air Pollution and Human Health," *Science* 169:723-733, August 1970.
13. Talbot Page, R. H. Harris, and S. S. Epstein, "Drinking Water and Cancer Mortality in Louisiana," *Science*, Vol. 193, 2 July 1976, pp. 55-57.
14. Waldbott, *Health Effects*.
15. *Ibid.*
16. W. W. Murdock ed., *Environment* (Sunderland, MA: Sinauer Associates, 1975), p. 281.
17. P. O. Cook, et al., "Asbestiform Amphibole Minerals: Detection and Measurement of High Concentration in Municipal Water Supplies," *Science*, 185: 853-855, September 1974.
18. L. J. Carter, "Pollution and Public Health: Taconite Case Poses Major Threat," *Science*, October 1974.
19. Epstein, *The Politics of Cancer*.
20. James F. Crow, "Chemical Risk to Future Generations," *Scientist and Citizen*, June-July 1968.
21. H. A. Schroeder and W. A. Vinton, "Hypertension Produced in Rats by Small Doses of Cadmium," *American Journal of Physiology*, Vol. 202:515-518, 1962.
22. G. W. Leeper, *Managing the Heavy Metals on the Land* (New York: Marcel Dekker, 1978), p. 56.
23. Luther J. Carter, "Michigan's PBB Incident: Chemical Mix-up Leads to Disaster," *Science*, Vol. 192, 16 April 1976, pp. 240-243.

THREE *Nature's Community*

1. Rena Corman, *Air Pollution Primer* (New York: American Lung Assoc., 1974).
2. Barry Commoner, *The Politics of Energy*, (New York: Alfred Knopf Publishers, 1979).
3. Kneeland A. Godfrey, Jr., "Land Treatment of Municipal Sewage," *Civil Engineering—ASCE*, September 1973).
4. S. S. Searle and C. F. Kirby, "Waste Into Wealth," *Water Spectrum*, Fall 1972.
5. John R. Sheaffer, "Pollution Control: Wastewater Irrigation," *DePaul Law Review*, Summer 1972.
6. Office of the Chief of Engineers, "Regional Wastewater Management Systems for the Chicago Metropolitan Area, Summary Report," March 1972.
7. S. M. King, "Fluid Wastes for More Fertile Farms," *Fertilizer Solutions*, May-June 1971.
8. Geoffrey Leeper, *Managing the Heavy Metals on the Land*, (New York: Marcel Dekker, 1978).
9. T. D. Hinesly, "Sludge Recycling: The Most Reasonable Choice?" *Water Spectrum*, January 1973.
10. Jeff Stansbury, "Of Human Waste and Human Folly," *The Living Wilderness*, Spring 1974.
11. Thomas Grubisich, "Radical Shifts Expected in U.S. Sewage Treatment," *Washington Post*, 30 April 1978.
12. Rene Dubos, *A God Within* (New York: Charles Scribner's Sons, 1972).
13. Alvin Toffler, *Future Shock* (New York: Random House, 1970).
14. *Ibid.*

FOUR *The Human Community*

1. William G. Pollard, "God and His Creation," *This Little Planet*, ed. Michael Hamilton (New York: Charles Scribner's Sons, 1970).
2. Joseph Sittler, Jr., "A Theology for Earth," *The Christian Scholar*, June 1954, pp. 367-374.
3. Alvin Toffler, *The Eco-spasm Report* (New York: Bantam Books, 1975).
4. Dennis L. Meadows, et al., *The Limits to Growth* (New York: Universe Books, 1972).
5. Paul R. Ehrlich, Anne E. Ehrlich, and John P. Holdren, *Human Ecology* (San Francisco: W. H. Freeman & Co., 1973).
6. Kenneth E. Boulding, "Commons and Community: The Idea of a Public," *Managing the Commons*, eds. G. Hardin and J. Boden (San Francisco: W. H. Freeman & Co., 1977).
7. Margaret Mead, "Community," *No Deposit—No Return*, ed. H. Johnson (Reading, MA: Addison-Wesley, 1969).
8. Lynn White, Jr., "The Historical Roots of our Ecological Crisis," *Science*, Vol. 155, March 1967, pp. 1203-1207.
9. *Ibid.*
10. Francis A. Schaeffer, *Pollution and the Death of Man* (Wheaton, IL: Tyndale House, 1970).
11. Rene Dubos, *A God Within* (New York: Charles Scribner's Sons, 1972).
12. "Deforestation and Disaster," *Time*, 22 May 1978.
13. Augustine, *City of God* (Garden City, NY: Doubleday-Image, 1958).
14. John Calvin, *Calvin's Institutes* (Grand Rapids: Wm. B. Eerdmans, 1949).
15. Erich Sauer, *The King of the Earth* (Grand Rapids: Wm. B. Eerdmans, 1962).
16. Inez Marks Lowdermilk, *Modern Israel: Fulfillment of Prophecy* (Berkeley: California Christian Committee for Israel, 1975).
17. Alastair I. MacKay, *Farming and Gardening in the Bible* (Emmaus, PA: Rodale Press, 1950).
18. Eric Rust, *Nature—Garden or Desert?* (Waco, TX: Word Books, 1971).
19. Frederick Elder, *Crisis in Eden* (Nashville: Abingdon Press, 1970).
20. Carl E. Armerding, "Biblical Perspectives on the Ecology Crisis," *Journal of the American Scientific Affiliation*, March 1973.
21. Francis A. Schaeffer, *Pollution and the Death of Man*, 1970.
22. Garret Hardin, "The Tragedy of the Commons," *Science*, Vol. 162, 1968.
23. Joseph Sittler, Jr., *"A Theology for Earth,"* June 1954.
24. Jerry H. Gill, *"The Ethics of Environment,"* The Reformed Journal, May 1978.

FIVE *Switching Power*

1. Lewis H. Lapham, "The Energy Debacle," *Harpers*, November 1974.
2. Peter Iseman, "The Arabian Ethos," *Harpers*, February 1978.
3. Edward Cowan, "World Oil Shortage Called Inevitable," *New York Times*, 17 May 1977.
4. Amory B. Lovins, *Soft Energy Paths* (Cambridge, MA: Ballinger Publishing, 1977).
5. Edward Delong, "Liquid Natural Gas: Savior Or Timebomb?" *Chicago Tribune*, 5 March 1978.
6. Paul R. Ehrlich, *et al., Ecoscience: Population, Resources, and Environment* (San Francisco: W. H. Freeman & Co., 1977).

7. Reinder Van Til, "Voices from the Mountains," *Reformed Journal*, November 1977.
8. Sean Toolan, "Cornfield Conflict," *Chicago Tribune*, 25 June 1978.
9. Terry Brown, "Oil, Wild Battle Looms," *Chicago Tribune*, 11 June 1978.
10. *Mineral Resources and the Environment* (Washington, D.C.: National Academy of Sciences, 1975).
11. Kenneth R. Sheets, "Some Second Thoughts about Coal," *U.S News & World Report*, 13 March 1978.
12. Lee Schipper, "Raising the Productivity of Energy Utilization," *Annual Review of Energy*, eds. Hollander and Simmons (Palo Alto, CA: Annual Reviews, 1976).
13. American Institute of Architects, *A Nation of Energy-Efficient Buildings by 1990* (Washington, D.C.: AIA, 1975).
14. C. Berg, "Conservation In Industry," *Science*, 12 April 1975.
15. Jet Propulsion Laboratory, *Should We Have a New Engine? An Automobile Power Systems Evaluation* (Pasadena, CA: JPL, 1975).
16. Committee on Nuclear and Alternate Energy Systems, John Gibbons, ed., "U.S. Energy Demand: Some Low-Energy Futures," *Science*, Vol. 200, 14 April 1978.
17. Edward Teller, "Nuclear Salvation," *Newsweek*, 17 May 1976.
18. Mary McGrory, "She Sees a Future in the Atom, *Washington Star*, 10 April 1978.
19. Alvin M. Weinberg, "Social Institutions and Nuclear Energy," *Science*, Vol. 177, 7 July 1972, pp. 27-34.
20. E. Flattau and J. Stansbury, "It Takes Energy to Produce Energy: The Net's the Thing," *Washington Monthly*, March 1974.
21. Howard T. Odum and Elizabeth C. Odum, *Energy Basis for Man and Nature* (New York: McGraw Hill Co., 1976).
22. David Comey, "Will Idle Capacity Kill Nuclear Power?" *Bulletin of Atomic Scientists*, Vol. 30, 1974, pp. 23-28.
23. Allan J. Mayer and William J. Cook, "A Boom Gone Bust," *Newsweek*, 31 October 1977.
24. F. Lundin, J. Lloyd, J. Smith, E. Archer, and D. Holaday, "Mortality of Uranium Miners in Relation to Radiation Exposure, Hardrock Mining, and Cigarette Smoking, 1950-1967," *Health Physics*, Vol. 16, 1969, pp. 571-578.
25. Casey Bukro, "A-Plant Paralyzed by Waste Disposal Dilemma," *Chicago Tribune*, 23 April 1978.
26. Casey Bukro, "Atom Foes Map Morris Showdown," *Chicago Tribune*, 14 May 1978.
27. Bill Richards, "Plutonium Taints Reservoir," *Manchester Guardian*, 3 April 1977.
28. Andrew Cockburn, "The Nuclear Disaster They Didn't Want to Tell You About," *Esquire*, 23 April 1978.
29. Casey Bukro, "Oak Ridge Reactor Breeds Controversy," *Chicago Tribune*, 18 April 1978.
30. Gerard Bonnot, "The Case of the Missing Uranium," *Atlas*, April 1977.
31. *Ibid.*
32. Hannes Alfven, "Fission Energy and Other Sources of Energy," *Science And Public Affairs*, January 1974.
33. David Gosling, "The Nuclear Power Controversy," *Atlas*, April 1977.
34. Barry Commoner, *The Politics of Energy* (New York: Alfred Knopf, 1979).
35. Denis Hayes, *The Solar Energy Timetable* (Washington, D.C.: Worldwatch Institute, 1978).

36. Ronald Kotulak, "Silent Energy Source Gains Importance," *Chicago Tribune*, 5 December 1977.
37. *Ibid.*

SIX *New Horizons*

1. R. R. Parizek, *et al.*, *Waste Water Renovation and Conservation, Penn State Studies 23*, Pennsylvania State University, 1967.
2. John R. Sheaffer, *Muskegon County Plan for Managing Waste Water*, John R. Sheaffer and Associates, Chicago, 1969.
3. U.S. House of Representatives, "Federal Water Pollution Control Act Amendments," Public Law 92-500, 1972.
4. U.S. House of Representatives, "Clean Water Act of 1977," Public Law 95-217.
5. David R. Zwick and Marcy Benstock, *Water Wasteland: Nader Task Force Report on Water Pollution* (Washington, DC: Center for Study of Responsive Law, 1971).
6. J. Brian Hyland, *A Report to the Committee on Appropriations, U.S. House of Representatives on the Environmental Protection Agency Construction Grants Program*, Washington, DC, 1979.
7. *Lubbock Christian College: Lubbock Land Treatment System Research and Demonstration Program*, Sheaffer & Roland, Inc.—Engineering Enterprises, Inc., Chicago, 1977.
8. *Northglenn Water Management Program*, Sheaffer &Roland, Inc., Chicago, 1977.
9. Pamphlet, "Career Development Center," Chicago Board of Education, 1978.
10. *Woodridge Environment School District 68: A School Conceived to Relate the Students to Their Environment*, Sheaffer & Roland, Inc., Chicago, 1976.
11. *Wastewater Management System, Itasca Center*, Sheaffer& Roland, Inc., Chicago, 1978.
12. John R. Sheaffer, "Stormwater for Fun and Profit," *Water Spectrum*, Fall, 1970.
13. U.S. Presidential Executive Order 11988, "Floodplain Management," May 1977.
14. *Living with a River in Suburbia*, Bauer Engineering, Inc., Chicago, 1972.

SEVEN *Life-styles That Count*

1. S. David Freeman, *Energy—The New Era* (New York: Random House, 1974).
2. Amitai Etzioni, "Choose We Must," Franklin Foundation Lecture Series, Georgia State University, 20 February 1979.
3. Greg Cailliet, Paulette Setzer, and Milton Love, *Everyman's Guide to Ecological Living* (New York: Macmillan, 1971).
4. Garrett DeBell, ed., *The Environmental Handbook* (New York: Ballantine Books, 1970).
5. G. Tyler Miller, Jr., *Living in the Environment: Concepts, Problems, and Alternatives* (Belmont, CA.: Wadsworth Publishing Company, 1975).
6. A. O. Tsui and D. J. Bogue, *Declining World Fertility: Trends, Causes, Implications* (Washington, DC: Population Reference Bureau Bulletin, October 1978).
7. E. F. Schumacher, *Small Is Beautiful* (New York: Harper & Row, 1973).

EIGHT *The Politics of Ecology*

1. Frances Moore Lappe and Joseph Collins, *Food First: Beyond the Myth of Scarcity* (New York: Houghton Mifflin, 1977).
2. Charles A. Job, "Basin Plan Aims at Lake Pollution," *Great Lakes Communicator*, Ann Arbor, Michigan, Vol. 9 (1), October 1978.
3. Jim Yoho and Virginia Scott, "You Can Help," *Illinois Environmental Council News*, Springfield, Illinois, Vol. 4 (8), December 1978.
4 John R. Sheaffer, "Pollution Control: Wastewater Irrigation, 1972," *The De Paul Law Review*, De Paul University, Chicago, Vol. 21 (4).
5. House Committee on Public Works, "Federal Water Pollution Control Amendments of 1972," H. R. Doc. No. 11896, 92nd Congress, 2nd Session 23 (1972).
6. "Clean Water Act of 1977," Public Law 95-217, 27 December 1977, 91 Stat. 1566.
7. *The Eighth Annual Report of the Council on Environmental Quality*, December 1977, U. S. Government Printing Office, Washington, DC.
8. Thomas C. Emmel, ed., *Global Perspectives on Ecology*, (Palo Alto, California: Mayfield Publishing Company, 1977).
9. *Ibid.*, pp 433-436.

NINE *The Blossoming Earth*

1. From the "Draft Policy Statement" of the American Clean Water Association, *ACWA News*, January-February 1979.

BIBLIOGRAPHY

The following books on various aspects of the environmental movement give further information or different perspectives.

Christian or Religious Perspectives
Derr, Thomas S. *Ecology and Human Need.* Philadelphia: Westminster Press, 1975.
Elder, Frederich, *Crisis in Eden.* Nashville: Abingdon Press, 1970.
Hatfield, Mark. *Between a Rock and a Hard Place.* Waco, TX: Word Incorporated Pocket Book Edition, 1977.
Jegen, Mary Evelyn and Manno, Bruno V. ed. *The Earth is the Lord's.* New York: Paulist Press, 1978.
Rust, Eric G. *Nature—Garden or Desert.* Waco, TX: Word Books, 1971.
Schaeffer, Francis. *Pollution and the Death of Man.* Wheaton, IL: Tyndale House Publishers, 1970.
Sider, Ronald J. *Rich Christians in an Age of Hunger.* Downers Grove, IL: InterVarsity Press, 1978.
Taylor, John V. *Enough is Enough.* London, England: SCM Press, 1975.

General Surveys on Environmental Issues.
Carson, Rachel. *Silent Spring.* Boston: Houghton Mifflin, 1951.
Commoner, Barry. *The Closing Circle.* New York: Alfred Knopf, 1971.
De Bell, Garrett. *The Environmental Handbook.* New York: Ballantine Books, Inc., 1970.
Dubos, Rene. *A God Within.* New York: Scribners, 1972.
Gardner, John W. *The Recovery of Confidence.* New York: W.W. Norton, 1970.
Hayes, Denis. *Rays of Hope.* New York: Worldwatch Institute, W.W. Norton & Co., 1977.
Leopold, Aldo. *A Sand County Almanac.* New York: Ballantine Books, Inc., 1970.
Schumacher, E.F. and Gillingham, P.N. *Good Work.* New York: Harper & Row, 1978.

Schumacher, E.F. *Small is Beautiful: Economics As If People Mattered.* New York: Harper & Row, 1973.

Selected Ecology Textbooks

Ehrlich, Paul R.; Ehrlich, Anne H.; and Holdren, John P. *Ecoscience.* San Francisco: W.H. Freeman & Co., 1977.

Miller. G. Tyler, Jr. *Living in the Environment.* Belmont, CA: Wadsworth Publishing Co., 1979.

Odum, Eugene P. *Fundamentals of Ecology.* Philadelphia, PA: W.B. Saunders & Co., 1971.

Wagner, Richard H. *Environment and Man.* New York: W.W. Norton & Co., 1978.

Environmental Topics

Commoner, Barry. *The Politics of Energy.* New York: Alfred Knopf, 1979.

Epstein, Samuel. *The Politics of Cancer.* San Francisco: Sierra Club Books, 1978.

Galbraith, J.K. *The Affluent Society.* Boston: Houghton Mifflin, 1958.

Lappe, Frances M. *Diet for a Small Planet.* New York: Ballantine, 1975.

Lappe, F.M. and Collings, Joseph. *Food First, Beyond the Myth of Scarcity.* Boston: Houghton Mifflin, 1977.

Leeper, Geoffrey. *Managing the Heavy Metals on the Land.* New York: Marcel Dekker, Inc., 1978.

Simon, Art. *Bread for the World.* Grand Rapids: Eerdmans Publishing Co., 1975.

Stevens, Leonard A. *Clean Water, Nature's Way to Stop Pollution.* New York: E.P. Dutton & Co., 1974.

INDEX

150